Rebecca L. Haller, HTM
Christine L. Kramer, BA
Editors

Horticultural Therapy Methods
Making Connections in Health Care, Human Service, and Community Programs

Pre-publication
REVIEWS,
COMMENTARIES,
EVALUATIONS . . .

"I was reminded of the enduring importance and meaning of horticulture throughout human history. The authors present horticulture and horticultural therapy as a wonderfully flexible medium that can engage people of all ages and abilities in learning, developing needed skills, and in making valued contributions to their communities. Those interested in or using horticulture for therapeutic, educational, or developmental purposes will find this book to be an excellent resource. With a view towards client-centered services and accountability, the book provides a foundation to build effective horticultural therapy services. Of particular value are the book's practical organization, the many useful examples of horticultural therapy in practice, and the strategically placed tips and resources. Extensive appendices provide the practitioner with detailed ideas and structure for horticulture activities that can benefit different client populations, followed by examples of documentation."

Karen C. Spencer, PhD, OTR
Associate Professor,
Occupational Therapy,
Colorado State University

More pre-publication
REVIEWS, COMMENTARIES, EVALUATIONS . . .

"The editors and authors have done a great job of providing professionals, students, and volunteers in the field of horticultural therapy a fantastic resource. The book is clear, concise, and logically organized, setting forth guidelines for the comprehensive practice of horticultural therapy from planning through implementation, evaluation, and documentation. A focus on purposeful practice with clear goals that address identified needs make this a must-read text for all horticultural therapy practitioners. Practices put forth are applicable to nearly any HT setting. Practical "Tips for Practice" are provided, making this an exceptionally useful book. The editors have assembled a wealth of information and resources for the appendix. *Horticultural Therapy Methods* truly is a comprehensive guide to best practices in horticultural therapy."

Jack Kerrigan, PhD
Assistant Professor, County
Extension Director,
Ohio State University Extension,
Cuyahoga County

"Kudos and best wishes to the editors for making this book happen! This is a detailed book designed to assist the reader with an in-depth study of the value of using horticulture as therapy. As a practicing horticultural therapist for 32 years, I found this book to be an ideal resource for the novice professional and seasoned therapist. Special chapters will enable you to use horticulture as a therapeutic tool for vocational, therapeutic, and social programs. The editors have brought together a plethora of resources and veteran professionals such as Charles Lewis and the Kaplans. These pioneers will stimulate and motivate the horticulturist, therapist, and volunteer in understanding the power of horticultural therapy to provide program strategies for client functioning, emotional healing, and health care enrichment. As editor Haller states 'One of the wonders of using horticulture as a therapeutic tool is how a connection with nature can affect someone deeply and profoundly.'"

Mitchell L. Hewson HTM LT RHCP
Author of the award-winning book
Horticulture As a Therapy

"In reading this book I am reminded of a statement Dr. Karl Menninger once said concerning what horticultural therapists needed for professional recognition: 'There are many ways (processes) to grow potatoes, but at the end of the season, a successful horticultural therapist will be able to dig a bushel of potatoes (product).' This text contains valuable information on the processes of planning and developing therapeutic horticultural activities, and also the procedures for documentation of the product of horticultural therapy intervention."

Richard H. Mattson, PhD, HTM
Professor, Horticultural Therapy,
Kansas State University

Horticultural Therapy Methods

Making Connections in Health Care, Human Service, and Community Programs

Haworth Series in Therapy & Human Development Through Horticulture

Rebecca L. Haller, HTM
Editor

Horticultural Therapy Methods: Making Connections in Health Care, Human Service, and Community Programs by Rebecca L. Haller and Christine L. Kramer

Titles of related interest

Horticulture as Therapy: Principles and Practice by Sharon P. Simson and Martha C. Straus

Generations Gardening Together: Sourcebook for Intergenerational Therapeutic Horticulture by Jean M. Larson and Mary H. Meyer

Horticultural Therapy Methods
Making Connections in Health Care, Human Service, and Community Programs

Rebecca L. Haller, HTM
Christine L. Kramer, BA
Editors

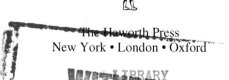

The Haworth Press
New York • London • Oxford

For more information on this book or to order, visit
http://www.haworthpress.com/store/product.asp?sku=5676

or call 1-800-HAWORTH (800-429-6784) in the United States and Canada
or (607) 722-5857 outside the United States and Canada

or contact orders@HaworthPress.com

The Haworth Press, Inc., 10 Alice Street, Binghamton, NY 13904–1580.

PUBLISHER'S NOTE
The development, preparation, and publication of this work has been undertaken with great care. However, the Publisher, employees, editors, and agents of The Haworth Press are not responsible for any errors contained herein or for consequences that may ensue from use of materials or information contained in this work. The Haworth Press is committed to the dissemination of ideas and information according to the highest standards of intellectual freedom and the free exchange of ideas. Statements made and opinions expressed in this publication do not necessarily reflect the views of the Publisher, Directors, management, or staff of The Haworth Press, Inc., or an endorsement by them.

Cover design by Kerry E. Mack.

Library of Congress Cataloging-in-Publication Data

Horticultural therapy methods : making connections in health care, human service, and community programs / Rebecca L. Haller,
    Christine L. Kramer, editors.
      p. cm.
    Includes bibliographical references and index.
    ISBN-13: 978-1-56022-325-2 (case : alk. paper)
    ISBN-10: 1-56022-325-1 (case : alk. paper)
    ISBN-13: 978-1-56022-326-9 (soft : alk. paper)
    ISBN-10: 1-56022-326-X (soft : alk. paper)
    1. Gardening—Therapeutic use. I. Haller, Rebecca L. II. Kramer, Christine L.
    [DNLM: 1. Gardening. 2. Occupational Therapy. WB 555 H822 2006]
RM735.7.G37H675 2006
615.8'515—dc22

            2006028220

To my husband, Jack Kramer, who is my rock and inspiration
of how to live life to its fullest, and to my children, Kateri
and Hannah, who keep me laughing.

Christine L. Kramer

To my parents, Deloris and Edward Haller, and my grandparents,
William and Sophia Thowe, who introduced me to the wonders
of the garden.

Rebecca L. Haller

CONTENTS

Chapter 4. Working with Program Participants: Techniques for Therapists, Trainers, and Program Facilitators

Karen L. Kennedy
Rebecca L. Haller

Sarah Sieradzki

ABOUT THE EDITORS

Rebecca L. Haller, HTM, has practiced and taught horticultural therapy since receiving an MS in horticultural therapy from Kansas State University in 1978. Currently the director of the Horticultural Therapy Institute in Denver, Colorado, Ms. Haller delivers workshops, teaches horticultural therapy classes in affiliation with Colorado State University, and provides consultation to new or developing programs. At Denver Botanic Gardens, she designed and taught a series of professional courses in horticultural therapy, managed the sensory garden, and created programs and access for people with disabilities. She established a vocational horticultural therapy program for adults with developmental disabilities in Glenwood Springs, Colorado, that is still thriving after more than twenty years in operation. She has served as president, secretary, and board member of the American Horticultural Therapy Association. In 2005, she received the Horticultural Therapy Award from the American Horticultural Society. She has authored articles and chapters in the following publications: *Horticulture as Therapy: Principles and Practice, Horticultural Therapy and the Older Adult Population, Towards a New Millennium in People-Plant Relationships, and The Public Garden.* She is also a member of the Thrive, the American Occupational Therapy Association, the American Therapeutic Recreation Association, and the American Horticultural Society.

Christine L. Kramer, BA, is the horticultural therapy program coordinator at the Horticultural Therapy Institute in Denver, Colorado. She and partner, Rebecca L. Haller, HTM, co-founded the institute in early 2002 and continue to offer accredited horticultural therapy education to students from across the United States and abroad. Previously, she worked as the program coordinator at the Denver Botanic Garden's horticultural therapy program. She has a BA in journalism from Metropolitan State College in Denver. Ms. Kramer has written for numerous publications including *OT Weekly, Mountain Plain and*

Horticultural Therapy Methods
© 2006 by The Haworth Press, Inc. All rights reserved.
doi:10.1300/5676_a

Garden, Green Thumb News, People-Plant Connection, AHTA News, GrowthPoint, The Community Gardener, Health and Gardens, Colorado Gardener, and *Our Sunday Visitor.* She was a reporter for the Denver Catholic Register newspaper for many years. She has received writing awards from the Society of Professional Journalists, the Catholic Press Association, and the Colorado Press Associations.

Contributors

Pamela A. Catlin, HTR, has worked in the field of horticultural therapy since 1976, and has been instrumental in establishing more than forty horticultural therapy programs in Arizona, Illinois, Oregon, and Washington. She is the director of horticultural therapy for Adult Care Services, Inc. located in Prescott, Arizona. She also does private contracting and is on the faculty for the Horticultural Therapy Institute in Denver, Colorado.

Karen L. Kennedy, HTR, is Manager of Wellness Programs at The Holden Arboretum, where she has coordinated the horticultural therapy program since 1986. The program includes training opportunities for professionals and student interns, and provides therapeutic services for people with chronic illness, traumatic brain injuries, mental retardation/developmental disabilities, people in physical rehabilitation, and older adults. She has actively worked with the American Horticultural Therapy Association (AHTA), serving on the board of directors, as a strategic area manager, secretary, and on work teams. Karen was the 1994 recipient of the Rhea McCandliss Professional Service Award from AHTA. She has given numerous presentations at both regional and national professional conferences and is a faculty member of the Horticultural Therapy Institute in Denver, Colorado.

Sarah G. Sieradzki, OTR/L, attended DePauw University and graduated from Indiana University in 1976 with a BS in Occupational Therapy. She was one of the first 200 occupational therapists to be licensed in Ohio. She worked in special education and has specialized in mental health since 1986. Her current title is clinical specialist in mental health occupational therapy at University Hospitals of Cleveland. Sarah has mentored students from at least six universities as a clinical supervisor. Beginning in 1992, Sarah actively pursued educa-

Horticultural Therapy Methods
© 2006 by The Haworth Press, Inc. All rights reserved.
doi:10.1300/5676_b

tion and experience in horticultural therapy, and has enthusiastically spoken about HT at local, state, and national conferences. She has co-led a horticultural therapy support group at Holden Arboretum with Karen Kennedy, HTR, since 1999. Sarah recently completed the HT certificate program at the Horticultural Therapy Institute in Denver.

Foreword

That horticultural therapy is maturing into a valued member of the international health care community is evident by the content and quality of this book. Having had the opportunity over the past forty-five years to observe and at times participate in the gradual evolution of horticultural therapy—from an activity conducted exclusively by volunteer gardeners, to a recognized therapeutic modality which includes trained, registered professionals—it is gratifying to see the emergence of such a clear, concise book that meets the needs of both professionals and volunteers. This book will be useful to educators in the health care arenas and the horticulture field as it provides an excellent framework for communicating the basics of horticultural therapy. At the same time, it leads readers to other resources for information and challenges them to meet the needs that exist for future growth of the field.

The authors are experienced practitioners and educators. As such, they have been able to identify and explore key issues that will help individuals coming into this field from health care or horticulture as professionals or volunteers to understand the theory, application, and impact of their work. This book is a valuable contribution to the growth of horticultural therapy and I would like to commend the authors for the dedication it has required to gain the skills and knowledge in horticultural therapy and to share them through their work.

Paula Diane Relf, PhD, HTM
Professor Emeritus, Virginia Tech University
Chair, People-Plant Council

Horticultural Therapy Methods
© 2006 by The Haworth Press, Inc. All rights reserved.
doi:10.1300/5676_c

Preface

This book was written to increase the value of using horticulture for therapy and human development by describing processes and techniques for practice. Based on many years of experience and teaching, the editors hope to encourage the competent use of these methods, which are widely accepted by allied professions. The manual is geared for repeated use as a reference and to help students, educators and those conducting horticultural therapy to provide effective and respected programs. It is also aimed at a variety of health care and human service professionals who use horticultural therapy, as well as community horticulture program leaders, and serious nonprofessionals such as volunteers and master gardeners.

For simplicity, throughout the text the term "client" is used to indicate the person served by the horticultural therapy program. It is intended to be a catchall term to refer to the consumer, inmate, resident, patient, trainee, student, or anyone who participates (for whom the program is directed). The reader is reminded that the client is first and foremost a unique person rather than a label. The use of "HT" as an abbreviation for "horticultural therapy" or "horticultural therapist" may be found in some areas of the text for brevity. The reader will not find the term "therapeutic horticulture" used here. Although this term has been used by the Thrive organization in the United Kingdom, and is gaining use in the United States, the editors believe that its use contributes to the challenges associated with describing horticultural therapy practice. Although the intention of those who use this term is to clarify HT practice and distinguish it from other beneficial uses of horticulture for human growth or wellness, its use may serve to muddy the waters rather than clarify them. Employers, donors, and laypersons may be further confused by the reversal of the two key terms. Thus, in this book, we have chosen to use "community horticulture" to describe those horticulture programs that exist to support human or community development. Leaders of these types of pro-

Horticultural Therapy Methods
© 2006 by The Haworth Press, Inc. All rights reserved.
doi:10.1300/5676_d

grams may not necessarily utilize all of the methods described in this book, yet will benefit from planning, leading, and documenting outcomes based on the procedures outlined.

The text is organized into five chapters. Chapter 1 sets the framework for discussion of the techniques in later sections. It also provides background about the shaping of the profession and how horticultural therapy is defined. Chapter 2 outlines the treatment process as used in horticultural therapy programs. Chapter 3 explains to the reader how to go about activity planning and developing horticultural therapy sessions. Chapter 4 talks about working with program participants and some techniques that are important for therapists, trainers and program facilitators. Last, Chapter 5 provides reasoning and directions for the documentation of treatment outcomes. The appendixes include forms and examples of documents used in the treatment process, as well as ideas for writing objectives and choosing relevant activities. It is hoped that readers will use these examples to create appropriate documents for the setting in which they practice.

At the end of each chapter, references are provided for the reader to further explore the topics covered. The manual focuses on skills involving treatment planning, client interaction, and activity selection—the nuts and bolts of practice. It does not attempt to be a comprehensive work on the field of horticultural therapy. Horticultural therapy also requires skills in program management, development, funding, garden design, enabling garden techniques, and facility management. More in-depth education and experience in understanding and serving populations encountered in HT programs are indispensable to maximize effectiveness. See the "Core Curriculum" of the American Horticultural Therapy Association (www.ahta.org) for a listing of these and other recommended topics for the continued study of horticultural therapy.

Readers should recognize that differing views exist within the profession on many of the topics presented. The perspectives given in this text represent each author's viewpoint and are not intended to represent the official views of the American Horticultural Therapy Association or any other professional organization.

The editors hope that the information in this book will boost the use of horticulture for therapy, rehabilitation, and wellness to continue to evolve in practice, scope, and recognition—so that more people may experience the benefits.

Acknowledgments

We are grateful to the many people who contributed to the production of this book, notably: the authors (Pam Catlin, Karen Kennedy, and Sarah Sieradzki) who provided their incredible experience and understanding of HT, Jay Rice, PhD, for his thoughtful contributions in the form of "practice perspectives," colleagues who offered suggestions for text edits (Terrell Kennet, Janet Laminack, and Sheila Taft), and the folks at Haworth Press for advice and interest in horticultural therapy.

Thank you to my son, Shawn Cremer, for sharing an enthusiastic relationship with nature and to Moss Cremer for his support and patience. Thank you to Christine Kramer for having the guts, resolve, and skills to work with me for so long—first in horticultural therapy at Denver Botanic Gardens, later as we began the Horticultural Therapy Institute in 2002, and most recently with this book. Without her encouragement and organization, we could not have pulled it off. I am also grateful to the students of horticultural therapy whom I have had the pleasure to teach over the years. They are an incredibly strong and determined lot. Thank you also to Richard H. Mattson, who pushed me to work with people with developmental disabilities and was my first mentor in HT; to Bruce Christensen, who gave me my first job in the field and inspired me to respect the clients served; and to the staff at Mountain Valley Developmental Services and Big Lakes Development Center for teaching me so much. Finally, my thanks go out to the clients with whom I have worked. They are the reason . . . .

Rebecca L. Haller

I met Rebecca Haller in 1994 and continue to be impressed with her dedication and drive to spread the gospel of horticultural therapy. Always uppermost in her mind is how to provide her students with excellence in HT education. I believe she has been successful and this book is just an extension of that strong commitment to help create leaders in this field. It's been my pleasure to learn from and work with such a devoted individual. I would also like to thank my former editor of the Denver Catholic Register, James Fielder. He was the finest editor I've ever had the privilege of working with in this field. Bless you, Jim. Last, I acknowledge my parents, Josephine and Jerry Capra, for always believing in my abilities, and my greatest cheerleader, my grandmother, Angelina Durando.

Christine L. Kramer

Chapter 1

The Framework

Rebecca L. Haller

INTRODUCTION

Whatever it is that calls the gardener to the garden, it is strong, primeval and infinitely rewarding.

Lauren Springer, *The Undaunted Garden*

Horticultural therapy offers positive and rewarding experiences for program participants, therapists, and those who come into contact with the growing environment. People involved in these programs intuitively know the many benefits and joys derived from connecting with nature. The attraction to engage in horticultural therapy activities may stem from a deeper "pull" in addition to the visible positive effects. Beneath the surface of this life-enhancing practice lie the conscious steps that are employed by the horticultural therapist to provide therapeutic programs. As an emerging profession, horticultural therapy continues to utilize the techniques practiced by many related health and human service fields—notably psychology, occupational therapy, vocational rehabilitation, social work, therapeutic recreation, and education. Much has been written about the reasoning and processes that are employed in these and other fields of human service. Theoretical bases for practice vary by setting, population, and therapeutic approach. However, the basic processes used are remarkably similar across disciplines. Horticultural therapy practitioners who adopt these accepted treatment procedures are able to positively impact the quality of their services and the profession as a whole.

Horticultural Therapy Methods
© 2006 by The Haworth Press, Inc. All rights reserved.
doi:10.1300/5676_01

This chapter provides a framework for discussion of the processes and techniques outlined in later sections of this manual. Included are systems and events that shape the horticultural therapy profession, a working definition of horticultural therapy, an overview of program types and the people served, and reflections on the significance of horticulture as a therapeutic medium.

SHAPING A PROFESSION

The practice of horticultural therapy has progressed from an 1800s' belief that working in the agricultural fields could benefit mental patients, to the use of gardening as activity and therapy for physical rehabilitation in the early 1900s, to the presence of many types of programming and settings in the 2000s. Entities that have influenced—and continue to shape—the profession of horticultural therapy include practitioners, educators, researchers, professional associations, volunteers, regulators, employers, as well as clients or consumers.

Practitioners

Since the mid-1900s, horticultural therapy has been used by mental health professionals, occupational therapists, physical rehabilitation specialists, vocational service providers, and others. In some settings, horticultural therapists work closely, or co-treat, with allied professionals to maximize outcomes and benefits to clients. The profession is characterized by a prevalence of practitioners who are willing to share information freely and who are open to new ideas and approaches. A healthy variety of program models result largely from the creativity and dedication of these diverse professionals.

Educators

Horticultural therapy education and training programs have led efforts to produce research, helped to define the profession through curricula content, and have been well-represented in establishing credentialing standards. Historically, and currently, most education programs are based in horticulture or plant science departments of colleges and universities, or in public gardens. Seldom are they found in human science or health care facilities. Future curricula with a higher proportion of human service and social science courses are needed to reflect the interdisciplinary competencies required of horticultural therapists who wish to practice in health care or social service arenas.

Researchers

Many benefits of horticultural therapy have been portrayed anecdotally in publications over the past fifty years. Research that documents outcomes and efficacy also exists, but is much less plentiful. The need for strong research as a base for the profession cannot be overstated. Practitioners, educators, and researchers from relevant disciplines must team together to carry out and publish research that employs sound social science methods. This is critical for future funding, employment, and excellence in practice.

Professional Associations

In the United States, the American Horticultural Therapy Association (AHTA) and its regional chapters have focused on information dissemination through publications, conferences, and networking. In order to advance the practice of horticultural therapy and expand employment options, future efforts must include strong advocacy and promotion of the profession to health care and human service providers as well as regulatory and insurance leaders (Haller, 2003).

Another important activity of AHTA is to create and manage a credentialing system—currently a voluntary professional registration system based on education, work experience, and professional activities. This system has recognized the scarcity of university degrees in horticultural therapy and has thus given credit for a wide array of education and experiences since its inception. Any new standards must take into account the diversity of horticultural therapy practice and educational options as well as the demands of the workplace and regulators.

Volunteers

Since the mid-1900s, volunteers have brought gardening activities to residents of prisons, hospitals, long-term care, and others—usually at no cost. Garden club members and master gardeners have been especially active in developing programs utilizing their training and experience in horticulture. Health care or human service education is seldom required, resulting in programs of varied therapeutic value. Distinguishing horticultural therapy programs and articulating a monetary value of volunteer services can help to eliminate the confusion that exists with administrators and potential employers. When volunteers serve as resources to programs led by trained horticultural therapists, optimum conditions for sustainability and effectiveness exist.

Regulators and Employers

Employers commonly look to supplemental funding bases to operate horticultural therapy programs. Private donations (of money, materials, and labor), self-earned income (from sales of plant products), and program grants are frequently vital to the provision of program funding. In health care, insurance companies regard horticultural therapy services as reimbursable only when they are framed within strict guidelines such as activity or co-treatment.

In order for the practice of horticultural therapy to be regarded as effective and fundable by administrators, insurance companies, and regulators, the following nationally coordinated actions must be taken: build a strong research base, apply standard treatment procedures to practice, develop a rigorous credentialing system, and advocate for the profession.

Clients

The people served in horticultural therapy programs play a role in how programs evolve. With a trend toward client-centered care, those involved may direct treatment plans and choose therapies, placements, and pursuits in which to be engaged. As more consumers understand the benefits of using horticulture as a tool for therapy and rehabilitation, they are more likely to choose those organizations that offer this service. For example, an elderly individual may prefer to live in an assisted living or long-term care facility that has a "gardening program" as an option for ongoing activity. This may give the organization a competitive edge and a motivation to continue or expand horticultural therapy programming.

HORTICULTURAL THERAPY DEFINED

As might be expected in a relatively new profession with diverse applications, a full spectrum of published definitions of horticultural therapy exists. Definitions encompass strict portrayals of horticultural therapy in health care terminology as well as those that broadly include any beneficial horticultural experience (Dorn and Relf, 1995). Recent authors have generally defined horticultural therapy more narrowly while using alternate terminology for the positive effects that gardening and passive garden exposure can have on the general population (Matsuo, 1992; Sempik, Aldridge, and Becker, 2003).

The following definition is presented to describe horticultural therapy in its many forms:

> Horticultural therapy is a professionally conducted client-centered treatment modality that utilizes horticulture activities to meet specific therapeutic or rehabilitative goals of its participants. The focus is to maximize social, cognitive, physical and/or psychological functioning and/or to enhance general health and wellness.

This definition includes three elements as described by Dorn and Relf (1995)—clients, goals, and treatment activities. They describe horticultural therapy as a practice that serves *defined client groups* (those with identified therapy or rehabilitation needs), that is *goal driven* (based on standard treatment procedures), and that uses the *cultivation of plants* as its primary treatment activities. These authors state the importance of the presence of all three elements to distinguish horticultural therapy from other types of garden interactions. In a graphic model of horticultural therapy, Mattson depicts the various interactions of a client, horticultural therapist, and plant during a horticultural therapy session (Mattson, 1982). A key element in this model, the trained horticultural therapist has the skills to use plants to facilitate the therapeutic process.

An adaptation of these earlier models shows the client as the central figure within the process of a horticultural therapy interaction (Figure 1.1). The client is both the receiver and initiator of the treatment process. In this model are four elements: client, goals, therapist, and plant. The client is the person being served—usually someone with an identified need for intervention to improve cognitive, emo-

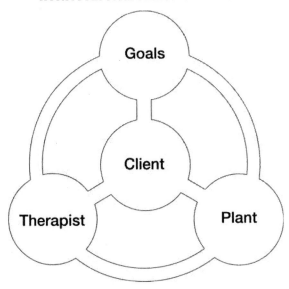

FIGURE 1.1. Practice Perspective: Horticultural Therapy Elements and Process

tional, physical, or social functioning. Goals are those treatment goals and objectives defined by the client and the treatment team. The therapist is the professional who is trained in the use of horticulture as a modality for therapy, rehabilitation, and wellness—the horticultural therapist. The term *plant* is used here to signify those garden- and plant-related activities and tasks used to provide therapeutic opportunities to the client. Pathways lead to and from each of the four elements. The therapist interacts with the client through the plant or with direct contact. Plant activities are chosen to meet the goals of the client. The client interacts with all other elements and is in the center of the diagram to illustrate a focus on the person in the horticultural therapy treatment process.

Note that horticultural therapy is based on purposeful horticultural activity or work—the cultivation of plants. Passive enjoyment of gardens and plants may be included as part of a program, but not to the exclusion of this active participation. Similarly, the guidance, judgment, and creativity of a trained therapist and stated treatment objectives are essential elements in horticultural therapy programming in order to maximize the benefits derived from the people-plant interaction.

It is anticipated that the definition presented here will help practitioners to know when and if their work should be identified as horticultural therapy. Programs that offer gardening activities without goals and treatment procedures, while beneficial, are not horticultural therapy. As an active participant in the horticultural therapy process, the therapist is responsible for planning treatment, developing sessions, interacting with the client, and recording outcomes or results.

Techniques and skills related to each of these elements are found throughout this manual. The training materials presented are based on the use of horticultural therapy as a treatment modality as defined earlier, with the intention that they will also enhance the skills of those who use horticulture as a tool for community development and other nonclinical human benefits. For example, a community garden coordinator who runs a program for at-risk youths will benefit from writing clear objectives and documenting outcomes. Although these and other community horticulture programs are not within a strict definition of horticultural therapy, they have in common the focus on human benefits rather than plant cultivation or physiology. Employing the basic methods described in this book will help community horticulture program leaders to clarify intentions, focus programming efforts, and produce measurable outcomes. In addition to more effective services, results include potential for new or continued funding and other support. See Exhibit 1.1.

EXHIBIT 1.1.
Practice Perspectives: Community Horticulture

Within this manual, the term *community horticulture* is used to describe those programs that use horticulture to improve quality of life through the development of individuals and communities. Examples of sites include community gardens and greenhouses, school gardens, urban greening sites, nature centers, healing gardens, and rural development projects. As with horticultural therapy programs, the purpose is to offer services that provide human benefits—including cognitive, psychological, physical, and social gains. In practice the boundaries between these programs and horticultural therapy are often blurred.

PROGRAM SETTINGS, TYPES, AND GOALS

Horticultural therapy programs vary in scope, setting, purpose, and treatment approaches. This diversity contributes to the appeal and affordability of horticultural therapy programs. From a few plants on a windowsill used for infrequent sessions to full-time programming and year-round growing facilities, programs can be customized to fit the needs and resources of each organization employing them.

Settings

Horticultural therapy programs are involved in health care, human service, health promotion or wellness, and community development. Settings include rehabilitation hospitals, mental health facilities, vocational training centers for people with developmental or other disabilities, correctional institutions, long-term care and assisted living facilities, schools, and community gardens. The purpose of programs can vary even within a setting. For example, in a rehabilitation setting, the horticultural therapy program may assist patients to regain functional physical or cognitive skills, help patients emotionally cope with loss or life-altering disabilities, provide guidance and exposure to a new leisure activity to increase quality of life, and/or may serve as a forum for improving social interactions and relationships. In a school setting, developmentally disabled students may participate in horticultural therapy sessions to improve work habits, social skills, or cognitive processes, whereas children with behavioral challenges may garden in order to learn socially appropriate behaviors, communication skills, and stress management strategies.

Types

Programs may be categorized into three types (Haller, 1998)— vocational, therapy, and social. Vocational programs seek to affect vocational outcomes of participants, improving job skills and employment. In therapeutic programs, the focus is on recovery from mental or physical illness or injury. A standard treatment approach is employed based on a medical model. A third type of horticultural therapy program, social or wellness, improves general health and/or quality of life of participants.

Goals

Within each of these program types, participants may work on specific goals in some or all of the developmental areas that are potentially impacted by horticultural therapy services, including cognitive (intellectual), emotional, social, and physical (Olszowy, 1978). The emphasis varies depending upon the program type and the individual clients served. Horticultural therapy is concerned with the needs and goals of the individual and whole person, rather than with a particular disease or disability.

Horticultural therapy utilizes horticulture activities to facilitate change in program participants. The purpose and focus of these activities varies by program type and by specific treatment objectives. Hagedorn (1995) identifies five focuses for the "applied use of activities." They are focus on product, process, competent performance, the individual interacting with others, and the individual interacting with the environment. Each of these approaches has merit and usefulness in horticultural therapy practice.

A *focus on product* or end result *of* the activity provides meaning and motivation to the participant. That a well-grown window box or harvest of cherry tomatoes is seen as a worthwhile product, results in therapeutic benefits to the client. The client is willing to participate and value the experience.

Therapeutic goals may be achieved by a *focus on the process* of doing an activity. Commonly used in horticultural therapy programs, the process itself can elicit improved mood or attention. By engaging in the process of gardening, a participant may experience flow (Hagedorn, 1995) or a sense of fascination (Kaplan and Kaplan, 1989) in which attention is effortless and complete, resulting in recovery from mental fatigue.

Through *competent performance* of an activity, the individual can begin to improve self-concepts and break negative cycles of real or perceived abilities and control. These inner changes can be catalysts for other therapeutic and developmental improvements. Through involvement in horticulture activities, a seed of change may be planted. For example, a patient may become more receptive to other therapeutic interventions and make progress in diverse areas. An inmate may perceive that he or she can succeed in school or work—leading to self-improvement in areas beyond the horticulture program.

Horticultural activities have the potential to be highly effective means for improving social *interaction*. Activities can be structured to encourage cooperation and communication among participants as well as to build relationships between the client and therapist. Barriers to communication are effortlessly reduced through shared garden work as shown in Exhibit 1.2.

Through activity, the participant interacts with his or her environment and consequently changes it in some way. In the horticulture environment the individual is also positively affected by this interaction with nature, with opportunities for growth, restoration and enjoyment (Kaplan and Kaplan, 1989).

More detailed descriptions of horticultural therapy settings, program types, and goals may be found in *Horticulture As Therapy: Principles and Practice* (Simson and Straus, 1998).

UNDERSTANDING THE PEOPLE SERVED

Understanding the populations who are served is essential in order to plan and implement appropriate and effective horticultural therapy treatment programs. Although it is important to know the cause, effect,

---

EXHIBIT 1.2. Tips for Practice: Facilitating Change

Goals to improve social interaction are often addressed during group activities, for example:

Goals: To improve communication skills, including making eye contact when speaking and initiating conversation

Task: To create a new summer garden

Process: The group meets to discuss the appropriate location for the garden, plants to include, and steps needed to prepare the site. The steps to prepare the site as well as the planting are divided and shared among the participants. This could occur over several sessions depending on time frame and level of the group. Each step in the process of creating a new garden offers many opportunities for participants to address their communication skill goals.

Contributed by K. Kennedy

and progression of many types of health conditions and disabilities, it is especially useful to recognize how conditions affect the individual's functioning and overall wellness. This knowledge impacts the reasoning, decisions, and interactions of the therapist on a daily basis.

For treatment planning, the horticultural therapist must know enough about a particular type of illness or disability, and, most important, about the individual, to develop activities or adaptations that are appropriate and accessible to the participant. Similarly, the organization or structure of the activity space must be based upon the needs and abilities of those served. This allows the client to access the space and the activity as well as experience appropriate challenges and problem-solving opportunities.

Safety considerations are also based on knowledge of the physical and cognitive skills and limitations of participants. The horticulture medium presents potential safety hazards through the use of tools, physical exertion, exposure to weather and sun, the cultivation of toxic plants, and the use of toxic or dangerous program materials. Informed choices must be made regarding what are or are not acceptable risks for each person served. In many cases, creativity is required to reduce hazards. See Table 1.1 for some examples of solutions to safety issues for an older adult with memory impairment.

TABLE 1.1. Examples from Practice: Safety Precautions

Issue	Solution
Susceptible to sunburn	• Ensure that the participant has sunscreen, long sleeves, hat, and sunglasses.
	• Schedule sessions when garden is not in full sun or work in shaded areas if possible.
May ingest non-food items	• Use nontoxic plants and program supplies.
Shuffles when walks	• Use hose reel to keep hose off walkways.
	• Use paved or hard-packed surfaces.
Disorientation or confusion	• Create pathways in a circular pattern with no dead ends.
Impaired judgment	• Limit access to sharp tools.

Contributed by K. Kennedy

Note: Safety precautions vary with program settings and needs of the participants so it is important to think through the issues and plan creative solutions.

Interactions between the therapist and client are improved if the therapist is familiar with key characteristics that influence communication with a particular population. For example, it is effective to use simple and adult-appropriate language with an adult with mental retardation. For someone with a hearing impairment, the therapist must speak clearly and look directly at the participant. In communicating with a person who has experienced a traumatic brain injury, it may be useful to provide written or pictorial directions to augment verbal instructions.

Practical descriptions of client populations may be found in "Part Two: Special Populations for Horticultural Therapy Practice" of *Horticulture As Therapy: Principles and Practice* (Simson and Straus, 1998). For information on diagnostic groupings, symptoms, functional limitations, and implications for treatment, see *Therapeutic Recreation and the Nature of Disabilities* (Mobily and MacNeil, 2002). Many of the major populations encountered in horticultural therapy are described in *PocketGuide to Assessment in Occupational Therapy* (Paul and Orchanian, 2003). Included are physical, psychosocial, geriatric, pediatric, and developmental diagnoses with summaries for each condition addressing etiology, symptoms, and outcomes.

In addition to this information, therapists need to understand the experience and social systems of the people served as described in Exhibit 1.3. Background information on general populations, combined with empathy and specific knowledge about each individual, are essential for effective horticultural therapy programming.

HORTICULTURE AS A THERAPEUTIC MODALITY

The use of horticulture as a therapeutic modality (or medium) offers numerous advantages, including a universal appeal, flexibility, and wide-ranging impact. What makes this medium special?

Encourages Human Growth

Charles A. Lewis devoted much of his life to understanding and communicating the positive effects of gardening on the gardener. In *Green Nature/Human Nature* (Lewis, 1996), he described several characteristics of gardening that encourage human growth and wellness.

EXHIBIT 1.3.
Practice Perspective: Understanding the People We Serve, An Ecological Empathetic Vantage Point

Heinz Kohut, the originator of self-psychology, suggested that empathy provides us with the experience of ourselves as a whole and vital self (Baker and Baker, 1987; Kohut et al., 1934). He defined empathy as the capacity to understand another's experience as well as the ability to mirror their experience back to them. Think of a time when someone you knew noted some quality they perceived in you. While you may have never thought of yourself in this way, you immediately felt the truth of what was said to you. In these moments of mirroring, we develop a more full view of ourselves as well as feel more significant and valued because we have been seen. Kohut suggests that throughout our lives our psychological well-being is enhanced by our empathic connections to others.

Empathy requires the ability to reflect upon our own life experience and to imagine the experience of others. In *Number Our Days*, noted cultural anthropologist Barbara Myerhoff (1978), describes imaginative identification, a process she utilized in studying aging with the members of a Jewish senior center in Venice, California.

> At various times, I consciously tried to heighten my awareness of the physical feeling state of the elderly by wearing stiff garden gloves to perform ordinary tasks, taking off my glasses, plugging my ears, slowing down my movements and sometimes by wearing the heaviest shoes I could find to the Center. ...Once by accident I stumbled slightly. The flash of terror I experienced was shocking. (Myerhoff, 1978, p.18)

In *The Ecology of Human Development*, Urie Bronfennbrenner (1979) suggests that empathy is aided by looking at the systems within which the individual develops, as well as those that currently impact the person's life. How did they grow up, who was available to meet their needs within their families, schools, churches, or communities? What was the level of communication between the family and others in their community? Were they isolated, alienated, or supported? If their parents worked, were they able to take time off to meet the needs of their children? What was the area like in which they grew up, physically, economically, and socially? How does the larger community view them? What private or public resources are committed to support or aid their participation in the community? When we work with someone who is

(continued)

(continued)

disabled, aging, or incarcerated, we may not get this information directly from him or her. However, secondary sources, such as research studies, books, films, etc., can help us begin to sense what it is like to walk in their shoes.

Contributed by Jay S. Rice, PhD, a psychotherapist in San Rafael, California, who integrates plants, nature, and healing ceremonies in his clinical work and a faculty member of the Horticultural Therapy Institute, Denver, Colorado.

These include

- enhanced pride and self-esteem resulting from the physical beauty created,
- deep personal involvement (individuals tend to become deeply engaged in the gardening process),
- requirements of patience and delayed gratification,
- awareness of natural forces and rhythms,
- interdependence or partnership between gardeners and the plants they tend (in this symbiotic relationship, the plants get tended; the gardener experiences a sense of purpose),
- inner peace based on the natural rhythms and dynamic stability of gardens, in contrast to modern schedules, fads, and threats, and
- opportunities for focused attention that provide rest from mental fatigue and worry.

Each of these innate or intrinsic characteristics of gardening lends something unique to the therapy experience.

Offers Restoration

Rachel and Stephen Kaplan (Kaplan and Kaplan, 1989) report that natural environments, including gardens, are restorative and are consistently preferred over many other environments. Restoration or recovery from mental fatigue results from contact with nature. In addition, gardener-reported experiences of enjoyment, relaxation, and lowered stress levels may ultimately affect physical health. Natural environments may also distract and allow the person to focus on something other than his or her current problem.

Addresses Innate Psychological Needs

Many authors have noted that gardening taps into a universal and innate psychological need for connections with plants and the earth (Lewis, 1996; Olszowy, 1978). In addition, gardening can provide many benefits:

- Food and other useful products
- Money from selling produce or saving on food costs
- Improved surroundings or environment
- Physiological conditioning
- Psychological benefits such as tranquility and healing and emotional health
- Mental growth
- Improved human relationships and communication
- Education in diverse topics (Matsuo, 1999).

Offers Versatility

The medium of horticulture is a flexible one that offers opportunities that cross cultures, ages, socioeconomic conditions, physical/mental/social/emotional conditions, and health status. Activities range from the many aspects of planning, preparing, planting, cultivating, harvesting, using products, and recycling garden wastes. The majority of tasks are easily adaptable to accommodate the abilities and challenges of diverse individuals.

Has Meaning and Purpose

Another key benefit of using horticulture is that it is a meaningful, purposeful activity that is motivating, normal, and tangible. In other words, it is "real." Many people enjoy the process as part of a normal healthy lifestyle. This normalcy and realness help break down barriers experienced by many program participants. See Table 1.2.

Gardening is a process that allows the gardener to be part of something bigger, to connect with nature, with the community and life. Horticultural therapy can provide experiences and insight to enhance

TABLE 1.2. Examples from Practice: Meaning and Purpose

Individual	Meaning
Older adult man in a nursing home	Gardening is a gender- and culturally appropriate activity and a continuation of a previous hobby.
MR/DD adults in a group home	Gardening provides produce for themselves and staff; it is also an activity and topic of conversation in common with community neighbors.
Outpatient mental health adults	Produce stand provides vocational and social skills by providing a valued commodity and opportunity to interact with the community.

Contributed by K. Kennedy

Note: Gardening is common to many people's experiences and family history. It also transcends barriers established by socioeconomic status, language ability and disabilities. It produces a product of value to others and enables the program participants to become the caregivers and the providers.

this sense of connection with natural rhythms. See Exhibit 1.4 for further information and examples of using nature ceremonies in horticultural therapy programs.

Impacts Others

Perhaps more than any other therapeutic medium, horticulture affects even those who do not receive horticultural therapy services. All who come in contact with horticultural therapy gardens and the resulting facility improvements can benefit from this passive interaction. This applies to visitors, administrators, and staff as well as residents who experience the contact with nearby nature on a daily basis.

Considering the importance of the horticulture environment and the role of plants in healing as described in Exhibit 1.5, it is imperative that horticultural therapists have a solid understanding of this medium as well as the necessary therapy skills to provide treatment. In horticulture, skills are needed in the areas of plant science, soils, indoor and outdoor gardening techniques, ornamental and food crops, enabling garden techniques, pest control, and organic growing methods. They may also need to know how to apply information about universal design, greenhouse or nursery management and therapeutic landscape principles. A wide repertoire of knowledge and gardening experience is

EXHIBIT 1.4. Practice Perspective:
Working with Nature: Ceremonies for Growth
and Understanding

When I first moved to San Francisco I came into a relationship with a 1968 VW Bus. One November day, I decided to try rebuilding the engine on my own. I bought a copy of *How to Keep Your Volkswagen Alive* and arranged to use my friend's driveway (Muir, 1969). November in the Bay Area happens to be the start of the rainy season. Over the course of the next two months, I spent much time on my back on wet and cold concrete. On a cellular level I was learning that it is best to make plans in harmony with the seasons. Of course farmers, gardeners, and horticultural therapists take direction from natural cycles as a matter of course. It would be easy for the horticultural therapist to take for granted aligning oneself with nature and miss how vital an offering this is to many people growing up in our modern and postmodern era. Humans, as part of the natural world, suffer when they do not recognize how they are affected by natural cycles.

A nature ceremony is one that aligns us with the season. For example, many people have difficulty with darker moods during the cold winter months. This condition is sometimes severe enough to be given the diagnosis of Seasonal Affective Disorder otherwise known as SAD (American Psychiatric Association, 2000). One treatment for this condition is exposure to light. Perhaps this is why Alaska Airlines has so many flights to Mexico in the winter. We might consider if there is something to be gained by the dark moods we experience during the winter months. After all, if we look at a deciduous tree in the middle of December, it would be easy to diagnosis it with depression. It has no fruits or leaves, and looks close to death. However, we know that the tree's response to the season is to draw its life force to its roots deep within the earth. Here it conserves energy and prepares for renewal in the spring. Perhaps, human suffering at this time of year is in part due to the fact that we often try to live our lives in the winter as if it is summer. We do not slow down and honor the pull of nature, rather we often push against it.

A nature ceremony teaches us that life is a process rather than a series of activities that are judged by whether they bring us immediately to our goals. In nature we learn patience by observing the stages of creation that proceed and follow fruition. In a nature ceremony we can help those we work with locate where they are in the cycle of an activity or their life journey (Eagle Star Blanket, 2004). We can plot out the stages: seed, germination, planting, new starts, flowering, broadcasting, death, compost, and awaiting rebirth. Through ceremonies we learn how to locate our human experience through metaphorical reflection and actual experience of our natural life cycle.

(continued)

(continued)

There are many forms of nature ceremonies that horticultural therapists can utilize. You can ask the people you are working with to plant an intention along with the planting of a seed. You can ask them to think of something they may wish to let go of or give away while they are weeding in the garden or dead heading. You can ask them to plant something at the start of a new endeavor and track their own development along with the growth of the plant. If you are working with individuals who feel isolated from their community or family, you can ask them to initiate an activity that teaches them about their connection to the earth. You might ask them to reflect upon how the sun, soil, plants, rocks, and air support their life? Where do they see themselves in the natural world? You can also help them cultivate images that support this feeling of interconnectedness. Ask them to imagine that the plant they are working with has a voice and can speak to them. What might this plant want to tell them? What might they want to say to the plant? Stephen and Rachael Kaplan explain that nature supports our use of indirect attention (Kaplan and Kaplan, 1989). Nature cultivates our capacity for reflection and for planning the course of our life. It allows us to see the big picture we often miss when engaged in our daily directed activities. From this vantage point, we are able to access wisdom regarding our life journey.

Contributed by J. S. Rice

EXHIBIT 1.5.
Practice Perspective: Relationship to Plants

Definition of Healing

It is important to recognize that our definition of healing influences our work with others. If we define healing as restoring someone to his or her original state of health, we may find that we often feel like we are failures or that our work is having little impact. This may particularly be the case if we are working with people with disabilities or those who have disorders brought on by the aging process. An alternative definition of healing is restoring one to the experience of wholeness and connection to the cycle of life. If our model of health implies physical perfection, we suffer needlessly. Our work in the garden and experiences in nature show us that things are rarely perfect in a national park picture way (Lopez, 1989). If we enter into the wilderness, we find

(continued)

(continued)

trees that were struck down by lightning valiantly growing toward the light. In our surroundings, we observe the breakdown and decay that supports new cycles of growth, as well as a teeming clutter that calls into question our heroic attempts to render life neat and orderly. If our model of wholeness connects us to this ongoing, vibrant, and sometimes chaotic natural impulse toward life that includes birth, life, death, and rebirth, we will help foster healing in those with whom we work. What call out for healing are the beliefs we hold that cordon off disease, disabilities, aging, and death from our understanding of life, nature, and community (Berry, 1983; Winkler, 2003).

How Cultivating Plants Teaches Us About Healing

When we face serious life threatening or disabling illnesses or diseases, we often become immobilized with fear and depression and withdraw from life. Our sense of self is affected emotionally, physically, mentally, and spiritually (Sacks, 1984). Cultivating a relationship with plants catalyzes movement in each of these aspects of our being (Rice, 2001). When people feel diminished in some way, they may feel they have lost the ability to offer something meaningful to the lives of others. Caring for plants provides physical confirmation of our ability to care for life (Rice, 1993). Physically caring for plants also provides an outlet for pent-up emotions. In essence, it is possible to give our emotions away to plants (Eagle Star Blanket, 2004). Physical activity also stimulates the production of endorphins that elevate our mood. Plants also provide us with an opportunity for reflection (Kaplan and Kaplan, 1989). As we work with them, we deepen our understanding of the life cycle. Plants teach us that every living being goes through periods of new beginnings, growth, emptiness, death, and rebirth. As we cultivate plants, they reciprocate by cultivating within us a deeper connection to the seasons of our life. In so doing, they inspire identification with the spirit of life that resides in all living things (Gibson, 1989; Rice, 2001).

Contributed by J. S. Rice

necessary to select the most effective treatment methods and activities in horticultural therapy practice.

SUMMARY

The field of horticultural therapy may be compared to an iceberg. When a client engages in a horticulture activity, the actions, and sometimes responses, are visible. We can see the garden flourish. We may be able to see attention, smiles, or calm expressions on the faces of participants. We may hear positive social interaction or expressions of joy. What are hidden beneath the surface are two very large aspects of horticultural therapy—the power of the relationships among the person, plant, and therapist, and the process and planning that accompanies the successful horticultural therapy session. On the surface it looks simple—underneath are complex reasoning and processes, as well as relationships that make this modality a compelling tool for working with people. As the horticultural therapist plans and sets up a gardening activity, based on the needs and skills of participants; a client or patient experiences the planting of a seed, digging in the soil, or tasting a ripe strawberry. A catalyst for positive change is thus set into motion.

These techniques and processes that underlie the visible parts of HT are described in later chapters and are intended to be applicable (with some adaptation) across the full range of programs. The horticultural therapy profession has evolved into one that offers varied treatment methods and serves a wide array of client groups. Programs are found in many types of health care, human service, and community settings

To provide a framework for succeeding chapters, this chapter has

- discussed some influences on the profession of horticultural therapy,
- defined horticultural therapy,
- noted the importance of understanding the people served by horticultural therapy programs, and
- described some important characteristics about the inherent value of using a horticulture medium for therapy and human development.

REFERENCES

American Psychiatric Association (2000). *Diagnostic and statistical manual of mental disorders DSM-IV-TR (text revision)*. Washington, DC: American Psychiatric Association.
Baker, H. S. and M. N. Baker (1987). Heinz Kohut's self psychology: An overview. *American Journal of Psychiatry*, 144, 1-9.

Berry, W. (1983). *A place on earth.* San Francisco: North Point Press.

Bronfenbrenner, U. (1979). *The ecology of human development: Experiments by nature and design.* Cambridge, MA: Harvard University Press.

Dorn, Sheri and Diane Relf (1995). Horticulture: Meeting the needs of special populations. *HortTechnology* 5(2): 94-103.

Eagle Star Blanket (2004). *Trail of prediction: Earth passage.* Conifer, CO: Eagle Dreams, Inc.

Gibson, Roberta (1989). *Home is the heart.* Rochester, VT: Bear and Co.

Hagedorn, Rosemary (1995). *Occupational therapy: Perspectives and processes.* Crawley, UK: Churchill Livingstone.

Haller, Rebecca L. (1998). Vocational, social, and therapeutic programs in horticulture. In Simson, Sharon P. and Martha C. Straus (Eds.), *Horticulture as therapy: Principles and practice* (pp. 43-68). Binghamton, NY: The Haworth Press.

Haller, Rebecca L. (2003). Advancing the practice of horticultural therapy. Unpublished presentation, AHTA Annual Conference, September 19, Victoria, British Columbia, Canada.

Kaplan, Rachel and Stephen Kaplan (1989). *The experience of nature: A psychological perspective.* New York: Cambridge University Press.

Kohut, Heinz, Arnold Goldberb, and Paul E. Stepansky. (1984). *How does analysis cure?* Chicago, IL: University of Chicago Press.

Lewis, Charles A. (1996). *Green nature/human nature: The meaning of plants in our lives.* Urbana, IL: University of Illinois Press.

Lopez, B. (1989). *Crossing open ground.* New York: Vintage Books.

Matsuo, Eisuke (1992). What we may learn through horticultural activity. In Relf, Diane, *The role of horticulture in human well-being and social development: A national symposium* (pp. 146-148). Portland, OR: Timber Press.

Matsuo, Eisuke (1999). What is "horticultural wellbeing"—in relation to "horticultural therapy"? In Burchett, M.D, J. Tarran, and R. Wood, eds., *Toward a new millennium in people-plant relationships.* Sydney, Australia: University of Technology, Sydney Printing Services.

Mattson, Richard H. (1982). A graphic definition of the horticultural therapy model. In Mattson, Richard H. and Joan Shoemaker (Eds.), *Defining horticulture as a therapeutic modality.* Manhattan, KS: Department of Horticulture, Kansas State University.

Mobily, Kenneth E. and Richard D. MacNeil (2002). *Therapeutic recreation and the nature of disabilities.* State College, PA: Venture Publishing, Inc.

Muir, J. (1969). *How to keep your Volkswagen alive: A manual of step-by-step procedures for the complete idiot.* Emeryville, CA: Avalon Travel Publications.

Myerhoff, B. (1978). *Number our days.* New York: Simon & Schuster.

Olszowy, Damon R. (1978). *Horticulture for the disabled and disadvantaged.* Springfield, IL: Charles C. Thomas.

Paul, Stanley and David P. Orchanian (2003). *PocketGuide to assessment in occupational therapy.* Clifton Park, NY: Delmar Learning.

Rice, J. S. (1993). Self-development and horticultural therapy in a jail setting. Unpublished doctoral dissertation, San Francisco School of Psychology, San Francisco.

Rice, J. S. (2001). A question of balance: Human-plant relations in the soul's journey. Unpublished presentation, AHTA Annual Conference.

Sacks, O. (1984). *A leg to stand on.* New York: Touchstone Books.

Sempik, Joe, Jo Aldridge, and Saul Becker (2003). *Social and therapeutic horticulture: Evidence and messages from research.* Leicestershire, UK: Media Services, Loughborough University.

Simson, Sharon P. and Martha C. Straus (Eds.) (1998). *Horticulture as therapy: Principles and practice.* Binghamton, NY: The Haworth Press.

Springer, Lauren (1994). *The undaunted garden: Planting for weather-resilient beauty.* Golden, CO: Fulcrum Publishing.

Winkler, G. (2003). *Magic of the ordinary: Recovering the shamanic in Judaism.* Berkeley, CA: North Atlantic Books.

Chapter 2

Goals and Treatment Planning: The Process

Rebecca L. Haller

INTRODUCTION

This chapter describes the process of treatment planning used in horticultural therapy programs. The focus is on treatment planning for individuals, a useful approach even when programs are delivered in a group setting. The aims of treatment in horticultural therapy, as in other therapies, are to help each client advance functioning in one or more areas and improve quality of life. In order to measure outcomes, therapists create plans and document achievements for each individual in the group.

Once someone is accepted or referred for horticultural therapy services, the treatment process begins. Prior to this point a determination is made regarding the appropriateness of horticultural therapy programming for this person. Depending upon the setting and systems in use, this decision is often made by a case manager, whose role typically includes referring clients for services, gathering information, coordinating and monitoring treatment and outcomes, and ending treatment or discharge.

TREATMENT TEAMS

Throughout the treatment process, the horticultural therapist may function as part of a treatment team that includes representatives from

Horticultural Therapy Methods
© 2006 by The Haworth Press, Inc. All rights reserved.
doi:10.1300/5676_02

various disciplines who are involved in the care or treatment of the person being served. The interdisciplinary team (IDT) is used in many health care, vocational, and educational programs. The team's make-up varies with the setting and individual, but generally includes key caregivers, therapists, family members, and the client himself or herself. Table 2.1 lists examples of additional IDT members. The IDT's responsibilities include implementation of all phases of the treatment process as well as periodic meetings—all with the aim to maximize the outcome for the client. By using the team approach, caregivers are encouraged to see the client as a whole person. He or she is known as a unique and complex individual with physical, social, psychological, cultural, and spiritual aspects. Increasingly in health care and human services, the client is central to the process and chooses care, goals, and treatment options. A team meeting that includes the client offers the therapist an opportunity to build a relationship with that client and to engage him or her in the therapy process. When the horticultural therapist is not part of the IDT, or when a team does not exist, it is important for the horticultural therapist to establish and use some form of regular communication with other parties or caregivers associated with the client. Access to case records and/or key information about the client is necessary for effective and safe programming.

TABLE 2.1. Examples from Practice: Interdisciplinary Team Members

Setting or Program Type	Possible Team Members (In addition to the client, family and HT)
Public schools	Teachers, nurse, speech therapist, occupational therapist, counselor, social worker
Vocational programs for people with developmental disabilities	Vocational rehabilitation specialist, psychologist, speech therapist, physical therapist, occupational therapist, nurse
Older adults	Physician/nurse, social worker, occupational therapist, activities professional
Mental health	Psychiatrist or counselor, social worker, therapeutic recreation specialist
Physical rehabilitation	Physician, physical therapist, occupational therapist, therapeutic recreation specialist, Hvocational rehabilitation specialist, speech therapist

THE PROCESS OF TREATMENT PLANNING

The team creates plans that are based on assessments of the client's functioning and the expected outcomes or achievements to be made. These may be called IEPs (Individual Education Plans), IPPs (Individual Program Plans), ITPs (Individual Treatment Plans), or other terms, depending upon the setting. (Further information about the components of an IEP may be found in Chapter 5, Exhibit 5.1. An example of a comprehensive individual plan is shown in Appendix E.) Plans are written using terminology that is behavioral or observable in order to clearly define the objectives and actions to be taken as well as to make it possible to measure the results. Plans must reflect the goals of the individuals being served.

As in other disciplines, in horticultural therapy, the treatment process follows a logical sequence from assessment to termination as outlined in Table 2.2.

Beginning with the assessment, or information-gathering phase, a foundation is laid for all interventions to come. This phase allows the horticultural therapist to build a solid program for the client, based on needs, desires, and abilities. Information may come from standardized assessment tools, interviews, observation, and/or case records. Through a process of careful review and discussion with the client, a problem or set of issues emerges. Particularly when many issues are present, choosing the one problem to be addressed requires care, skill, and thoughtfulness on the part of the horticultural therapist and other team members. It may be helpful to create lists and narrow the

TABLE 2.2. Tips for Practice: The Treatment Process

Phase	Actions
Assessment	Information gathering and analysis
Goal identification	Prioritization of problems, needs, and aims
Action plan (Intervention plan)	Creation of objectives, actions, and means for measurement
Intervention	The plan into action, horticultural therapy activities
Documentation	Recording of actions (occurs along with intervention phase)
Revision	Ongoing review of results and modification of plan
Termination	Discontinuation of treatment, reporting results

focus through a process of elimination (Buettner and Martin, 1995). Begin by listing needs and problems that have been expressed. Also list the strengths and interests of the client. With the latter in mind, begin to eliminate those problems on the first list that are

- low priorities for the client (or even unacknowledged by him or her as problems),
- relatively irrelevant to the client's quality of life,
- impractical to address in the horticultural therapy setting or program, or
- not likely to be successfully improved.

This process, along with communication with the client and caregivers, usually narrows the list to a few priorities. In general it is recommended that clients work on no more than three goals at a time. The more focused the treatment, the more likely goals will be met with measurable outcomes. A treatment focus does not eliminate the need to constantly be aware of the person as an entire, whole human being. He or she is not just the "problem" being addressed. In all interactions and horticultural therapy sessions, the therapist is mindful that each person and situation is unique.

The client's aims, concerns, and dreams must be considered. Even if a dream seems unrealistic at first, look for intermediate goals or steps that move toward that aspiration. It may be, in fact, possible to reach a dream or modify it to be achievable. Furthermore, a client whose wishes are respected will be more motivated to work on the steps involved in therapy.

Once the problems have been prioritized, goals to address them should be clearly stated. Goals are generally long-term and articulate something to be achieved through several steps or series of efforts. The following are examples of goals:

- Obtain independent employment in the community.
- Garden independently at home following a spinal cord injury.
- Maintain friendships within an assisted-living setting.
- Manage anger effectively on the school playground.

The next step in the process of treatment is an action plan that includes detailed objectives or short-term goals, actions to be taken, and plans for how to measure the outcomes. The action plan allows all parties involved to know what specific behaviors are desired, what actions and interactions will take place to facilitate their accomplishment and how the progress or results will be recorded. The therapist and treatment team uses a problem-solving approach to formulate ways to alleviate the issues facing the client. The following list defines the components of the action plan.

- *Objective:* A precisely worded statement of what is to be achieved within a short timeframe
- *Methods:* Who will do what, how, under what circumstances, where, when, and how often; includes what the therapist will do, the strategies for intervention
- *Criteria:* Quantifies achievement of objective, e.g., how much, how many, how often, how well. (The criteria may be included in the objective itself.)
- *Documentation:* Outlines who will measure, what will be measured, how often, and where or how it is recorded

As a key part of the action plan, the objectives are clearly written, detailed descriptions of what is to be achieved within a short specific timeframe. They are written behaviorally so that progress can be measured, and include specific criteria for achievement. Objectives also include information about what, by whom and how the actions will be carried out, referred to as "methods."

What makes a good objective? An essential characteristic is that the objective logically leads to long-range goal attainment. The team, including the client, must be thoughtful about its selection. Similarly it must be achievable and realistic under the circumstances. It must be precisely worded to avoid confusion and disagreement. It must be measurable, so that outcomes can be documented accurately. And, the client must agree to participate. Examples of objectives (based on the long-range goals stated earlier) are shown in Table 2.3.

In order for objectives to be clear to all parties involved and to facilitate the recording of outcomes, they should be written following the guidelines in Exhibit 2.1. Further directions for writing effective goals are discussed in Chapter 5.

The next step in the treatment process is intervention—carrying out the plan. This is when the horticultural therapy takes place— when the horticultural therapist, client, and plants interact with the

TABLE 2.3. Examples from Practice: Objectives

Long-Range Goals	Example Objectives
Obtain independent employment in the community.	Increase speed (while maintaining accuracy) on specified greenhouse tasks by 20 percent of current baseline rate for two consecutive weeks.
Garden independently at home following a spinal cord injury.	Use extended-handle gardening tools to plant bedding plants in prepared raised beds without assistance for three sessions.
Maintain friendships within an assisted living setting.	Initiate a casual conversation with at least one "garden club" member during each session for one month.
Manage anger effectively on the school grounds.	Control anger on the playground at recess by independently walking to and sitting quietly in the garden, away from situations and people that elicit outbursts, 90 percent of the time for four consecutive weeks.

purpose of reaching stated objectives. For most horticultural therapists and clients, this is the focus—the part of the process that is the most rewarding and enjoyable. Program and activity planning may take place at this stage, always based on treatment plans and objectives for the individuals involved. Chapters 3 and 4 provide more detail about developing horticultural therapy sessions and working with program participants. During the course of engaging in horticultural activities with participants, surprising behaviors may emerge as connections and relationships are built or strengthened. One of the wonders of using horticulture as a therapeutic tool is how a connection with nature can affect someone deeply and profoundly. Therefore, the horticultural therapist must be constantly alert for opportunities to enhance the therapeutic experience for each client served as well as carry out the action plans as written.

Concurrently with intervention, behaviors specified in the action plan are documented. Records are kept according to schedule and are used to show progress (or lack of progress) toward stated objectives. What is recorded depends upon the objectives in place. The agency or institution where the horticultural therapy occurs often dictates which forms or procedures are used. The important point is that objectives and documentation are aligned. In other words, the data recorded

EXHIBIT 2.1.
Tips for Practice: Writing Measurable Objectives

For an objective to be functional it should:

1. State the *desired behavior* or performance. Indicate what the client will do, what behaviors are expected. Include an action verb. Focus on one behavior per objective.
2. Identify *conditions* under which the behavior is to be performed. This could include where, or with what support, the action is performed.
3. Include *criteria* or standards to measure performance. Identify the criteria for acceptable performance, such as for how long, with what accuracy, how much, how well, etc. (Criteria may be included in the objective statement itself or listed separately on the action plan.)

The objective should be understandable by others, clearly specifying the desired behaviors. It should not include information about what the therapist will do, unless some form of support is included in the conditions. This information is part of an action plan or strategies for intervention, but not included in the objective itself.

must measure progress toward the objective. In some instances additional program information may be regularly noted, such as attendance, productivity, or notes about the process and activity. Documentation should reflect observable behaviors rather than subjective impressions or inferences. They should be kept confidential and only discussed with parties that are authorized to access that information—such as the treatment team and the client. Documentation is discussed further in Chapter 5.

Throughout the intervention stage of the treatment process, the therapist must monitor progress of the client. Using good documentation as a basis, the therapist notes achievements and/or lack of advancement and evaluates the objectives and action plan. In the revision phase, it may be necessary to modify target dates, criteria, methods, or even the objectives themselves to ensure growth and development. The client may progress faster or slower than anticipated, may have reached a ceiling of achievement or may have experienced other influences that affect advancement. If objectives have been met, the client should move to the next step as appropriate. For example, other steps toward the long-range goal "obtain independent employment in the community," might be to "begin work on own initiative" or to "get to the work site independently." The agency may specify procedures regarding revision of objectives or action plans. In some cases, the therapist involved makes simple adjustments routinely. In others, the IDT must meet to consider the changes or a case manager must approve them.

The final phase of the treatment process is termination or discharge. Horticultural therapy services may end for various reasons, including discharge of the client from the care setting, ending of a seasonal program, client progress (or regress) to a new program, goal achievement, or desire of the client. Again, procedures for ending horticultural therapy participation follow those of the employing agency. In general, the therapist helps to prepare the client for the transition, reviews documentation, and may write a summary report that includes objectives, achievements, and recommendations for other services, as needed.

SUMMARY

Horticultural therapy is practiced in diverse settings, under many different circumstances. This chapter outlined a process for providing horticultural therapy services that can produce measurable outcomes for program participants. In some settings, therapists need to modify this process in order to fit in with existing methods. In summary, the horticultural therapy programming should be based on prioritized needs or problems, be goal driven, be articulated clearly, and use documentation methods that show measurable results—all within the framework of the organization with which it is associated. This will allow the horticultural therapist to serve the clients well and to communicate results in a professional manner.

SUGGESTED READINGS

Austin, David R. (1991). *Therapeutic recreation: Processes and techniques*. Champaign, IL: Sagamore Publishing.

Borcherding, Sherry (2000). *Documentation manual for writing soap notes in occupational therapy.* Thorofare, NJ: Slack, Inc.

Buettner, Linda and Shelley L. Martin (1995). *Therapeutic recreation in the nursing home.* State College, PA: Venture Publishing, State College, PA.

Davis, William B., Katie E. Gfeller, and Michael H. Thaut (1992). *An introduction to music therapy: Theory and practice.* Dubuque, IA: Wm. C. Brown Publishers.

Hagedorn, Rosemary (1995). *Occupational therapy: Perspectives and processes.* New York: Churchill Livingstone.

Hagedorn, Rosemary (2000). *Tools for practice in occupational therapy: A structured approach to core skills and processes.* New York: Churchhill Livingstone.

Hogberg, Penny and Mary Johnson (1994). *Reference manual for writing rehabilitation therapy treatment plans.* State College, PA: Venture Publishing.

Mager, Robert F. (1997). *Preparing instructional objectives.* Atlanta, GA: The Center for Effective Performance.

Ozer, Mark, Otto D. Payton, and Craig E. Nelson (2000). *Treatment planning for rehabilitation: A patient-centered approach.* New York: McGraw-Hill.

Zandstra, Patricia J. (1988). A systematic approach to horticultural therapy. *Journal of Therapeutic Horticulture* III:15-24.

Chapter 3

Activity Planning: Developing Horticultural Therapy Sessions

Pamela A. Catlin

INTRODUCTION

Horticultural therapy facilitates connection between people and plants—between people and nature. This connection is the essence of how and why gardening activities are such powerful catalysts for positive human development. The garden provides people with a multitude of opportunities for seeing themselves, and the world, from a new perspective. Its associated activities can be used for consciously enhancing spiritual connections and growth. As activities are chosen, it is imperative that therapists look first to the garden to find those natural activities that enhance this relationship.

With this in mind, this chapter explores many of the variables that go into developing effective horticultural therapy activities, including

- goals addressed in the horticultural therapy program,
- selecting activities or tasks to meet therapeutic goals,
- additional considerations for activity selection,
- planning for seasonality, and
- resources for session ideas.

Horticultural Therapy Methods
© 2006 by The Haworth Press, Inc. All rights reserved.
doi:10.1300/5676_03

HORTICULTURE AS A VEHICLE
TO MEET WIDE-RANGING GOALS

On account of its broad ranging appeal and varying levels of skills required, horticulture lends itself to serve the needs of many diverse populations and ability levels. Some of the primary goal areas addressed in successful horticultural therapy programs are

- physical
- cognitive
- sensory stimulation
- emotional
- interpersonal/social
- community integration

Physical

Some of the physical benefits derived from a horticultural therapy program are obvious to the casual observer. Being out in a beautiful garden environment in the fresh air is most likely what comes to mind. However, many other physical goals and objectives can be met through the horticultural therapy program. Examples are maintenance or improvement of fine motor skills, gross motor skills, standing/balance and endurance, mobility, range of motion, and strength.

Often therapists will find participants willing to pursue physical therapy goals with greater ease in the garden than in the therapy room. The familiarity that a garden can bring, the sensory stimulation that is part of being surrounded by and working with plants, the distraction and focus of working with plants and the reward of the beauty and harvest—all can play a part in this process.

Cognitive

Cognitive benefits can be seen in such areas as

- speech and vocabulary,
- memory,
- learning new skills,
- sequencing,
- following directions,

- problem solving, and
- attention to task.

Incorporating information about the plants being used, such as the names of the plants and where they come from geographically, are ways to enhance a program. Orientation to time and place can be facilitated through the horticultural therapy program by incorporating special holidays and seasons as well as "checking in" about the progress on planting projects to give participants a gauge for time. Simple tasks such as writing one's name, date, and plant name on a label can accomplish several cognitive goals in one aspect of an activity.

Sensory Stimulation

The horticultural therapy program can be an excellent avenue for providing sensory stimulation. Often a required component of activity programs, sensory stimulation can be met through the use of fragrant flowers and herbs, plants with interesting textures and tastes, and specific color combinations. Sometimes overlooked, the auditory senses can be awakened by designing the garden to have trees and grasses that create subtle sounds or by using a water meter indoors that makes audible clicking sounds when plants are wet. The goal of sensory stimulation can be tied in with both cognitive and emotional goals when thinking of activities to be used. For example, a time for sharing the fragrance and taste of various herbs could stimulate a client's childhood memory of collecting herbs for his or her grandmother. More detail about responses to fragrance is shown in Exhibit 3.1.

Emotional

The horticultural therapy program can help create a stronger sense of self-esteem for participants. In situations where all too often choices and responsibilities are taken away, the well-run horticultural therapy program can provide a safe way to incorporate those functions back into people's lives. In settings where individuals are most often in the care-receiver role, the horticultural therapy program can provide an outlet for the participant to give care to the plants and to other people as well. Esteem building can take place through the un-

EXHIBIT 3.1.
Tips for Practice: Fragrant Responses

Fragrance can be a tool used with great impact in horticultural therapy programs. But it is important to note that everyone's reaction to fragrance is specific to him or her. Receptors in the nose are uniquely arranged, which is one of the reasons one person can detect more of some components in a fragrance than the next person and why we each experience fragrances differently. Ask a group to describe what they smell when presented with a lemon rose-scented geranium. Some will smell the rose, others will smell the lemon predominantly, and still others will detect other fragrances.

Cultural background can impact whether a smell is perceived as good or unpleasant. Cultural experience with certain foods for example, affects whether one enjoys or "turns up one's nose" to culinary herbs such as rosemary, oregano, or sage.

Fragrance also has the power to evoke powerful emotions tied to memories of specific times, places, and people. While the therapist usually intends happy, pleasant, and comforting associations, the emotional response can be either positive or negative. It is important to be prepared when the associations are unpleasant, uncomfortable, or sad. A gardenia flower might be associated with the pleasant memories of a high school dance, or it might remind someone of their recently deceased husband who gave them a gardenia corsage every Mother's Day. When this happens, the horticultural therapist should respond compassionately and allow the person to express his or her feelings about the memory.

Sometimes the association is linked to an experience. Preference for a particular plant or flower relates to the individual's perception of the situation. A plant that illustrates this rather dramatically is *Plectranthus* "Vick's Plant." It generally induces positive feelings for those who remember being comforted by their mother with a menthol rub while they were sick. Others associate the smell with feeling miserable and don't like the plant at all. It is helpful to have a variety of choices so that someone can choose a plant or flower with positive associations.

Contributed by K. Kennedy

limited opportunities for creative expression offered in a diverse horticultural therapy setting. Horticulture activities, more than many other hobbies, are often appropriate for any age and can easily be adapted to any ability level. Well-planned and implemented age-appropriate horticultural activities result in enhanced self-esteem.

The horticultural therapist can structure a program to provide emotional benefits in the areas of anger/aggression management. Physical activities can utilize the energy found in the anger or aggression and channel it into productive directions. Examples of successful activities in this area are digging to prepare the garden, weeding, raking leaves, mixing soil, or washing pots.

The emotional benefit of an improved outlook on life may be fostered through ongoing horticultural activities in which individuals are encouraged to look for signs of weekly, if not daily, growth and change. The cycle of propagating garden plants from seed is an example of this process. One's focus can evolve—starting with the needs and changes of the seedlings, to planting in the garden through harvesting. Such projects can provide those involved with a sense of wonder and purpose in life. The garden setting also enhances a spiritual connection through a sense of connectedness to nature and a witnessing of the cycles of life. See Exhibit 3.2. In the wellness arena, the emotionally rejuvenating link that nature provides for most people can be a stabilizing force in a rushed, stressful work environment.

Interpersonal/Social

The horticultural therapy program can provide social benefits in a number of ways. The simplest is the socialization that naturally occurs when participants are brought together for a group session.

EXHIBIT 3.2.
Practice Perspective: Spiritual Connection

The horticultural therapy program is an ideal setting for individuals to explore their own spiritual connection, and this can be profoundly transformational. Spirituality, not to be confused with organized religion, has to do with feeling a sense of connection with one's surroundings, with others, and with the natural cycles of life. For as long as there is recorded history, the spiritual connection between plants and people has appeared in art of all forms. In the fast-paced, man-made world or the typical institutional world, life can seem out of balance. There is an order in nature. The symbiotic relationship between the birds, butterflies, bees, and flowers or the perfection in a flower can help an individual to regain a sense of being at peace and renew a sense of awe and interest for the world of which they are a part.

There are many anecdotal reports of individuals with histories of being withdrawn who begin speaking and/or participating in a project when shown a blooming plant. On more subtle levels, socialization can take place through such things as

- enhancing family visits through sharing garden happenings,
- creating projects in a session,
- conversing with other people while weeding,
- sharing with others the bounty of the garden, and
- cooperating with others in gardening projects.

Another facet of social benefits is recreation or leisure skills. Every person, regardless of age or ability, needs a recreational outlet. Gardening, whether a new skill or a long-time favorite hobby, is an excellent leisure time activity. It helps motivate people to move out of a sedentary lifestyle. Gardening can be done by oneself or cooperatively with others and it helps to provide a healthier diet as well as add to the esthetics of a home or facility. Working with plants as a leisure skill is a common link with many other "nondisabled" individuals, reducing the stigmas that many people with disabilities might encounter in their personal leisure lives.

Community Integration

Because gardening (indoor and outdoor) is one of the top leisure activities of the general public, many opportunities exist for clients to integrate into the community through horticulture in a recreational sense. These include tours of local horticultural facilities, involvement with local garden clubs or participation in county fairs and community gardens.

Horticulture may be able to provide job opportunities within the community because green industry employers (nurseries, landscapers, etc.) need people with basic horticulture skills, such as watering, planting, and basic garden maintenance. For example, many programs serving at-risk youths, individuals with developmental disabilities or traumatic brain injuries, and those who are incarcerated provide on-the-job training, working directly with clients in the job setting.

Ideas and activities that may be used to address some of the overall goals discussed in this section may be found in Appendixes A and D.

ACTIVITY OR TASK SELECTION
TO MEET TREATMENT OBJECTIVES

In order to select horticulture activities that will be effective and advance each client closer to his or her goals, take into account the following:

- The *activity is a tool* for treatment.
- The *type of program* influences the selection of activities or tasks.
- *Treatment issues/goals* are paramount considerations for choosing what to do.
- Participants each have *varied backgrounds,* skills, and interests.
- A *situational assessment* may yield information to drive future session content.
- *Program continuity* helps connect sessions for enhanced outcomes.

Beyond Activity

Horticultural therapy is often activity based, yet it is of vital importance to remember that the gardening project is only a tool in the therapeutic process, not the end goal. Often a session developed around sensory stimulation and a greater awareness of one's surroundings can be more effective than a gardening project with many steps that results in something to carry out the door. There may be times in specific settings when an activity is used for diversion, but far more often the horticultural therapy activity is devised to assist individuals to achieve specified goals.

When therapists ask patients to expend energy in therapeutic groups, they need to have clear ideas about what makes an activity both appealing and useful (Borg and Bruce, 1991). The activity should be viewed as therapeutic by the therapist using it, and should be tied to an acceptable therapeutic theory and base. When therapists are fully engaged with the activity, they communicate that the activity is worthwhile and can lead to goal attainment. Desirable characteristics of an activity are that it

- is purposeful,
- is interesting,
- offers a chance to take responsibility,
- necessitates an investment of energy,
- can be task analyzed (as described in Chapter 4),
- is balanced between demands and patients abilities, and
- has a beginning and end.

Type of Program or Facility

The type of program or facility plays a major role in determining the selection of an activity or task. Following is a listing of the many types of facilities where horticultural therapy is used and examples of the programs typically found in those settings.

Vocational Organizations

Vocational programs might have a horticultural therapy program with an emphasis on training people in specific horticultural job skills such as potting, transplanting, watering, fertilizing, etc. The horticultural setting often is also used to provide individuals with basic job training including development of social skills. In addition, there is a trend for these programs to provide opportunities for learning new leisure skills, in which case a more recreational and social approach to activity selection would be necessary. See Exhibit 3.3 for an example of how vocational programs may address the needs of the whole person.

Long-Term Care

In the long-term care setting a recreational/social program is often utilized. Activities ideally would address general goals such as increased socialization, sensory stimulation, and cooperation with others. A therapeutic emphasis can also be implemented by selecting activities that address goals such as improved/maintained motor skills, cognitive skills, or orientation to time and place. This focus can take place either in the group setting or individual sessions. In long-term care, maintenance of skills is usually the goal. There often are, however, individuals who might reside at the facility for a brief period of rehabilitation. In this case, a focus on improvement of skills would

EXHIBIT 3.3.
Examples from Practice:
Goal Areas in Vocational Program

While a program may have primary target goals, it is not uncommon for a program to be able to address goals in many different areas over the course of time. This is particularly true of long-term treatment programs. For example: within a day treatment pre-vocational program for adults with traumatic brain injury, a garden area is maintained containing plants to be harvested for a fall herb vinegar sale. The following goal areas may be addressed:

- *Physical:* balance, coordination, and endurance
- *Cognitive:* practicing compensatory strategies for memory, planning and organizational skills, writing skills
- *Emotional:* anger management, impulse control, stress management, sense of accomplishment and success; self esteem and nurturing opportunities
- *Spiritual:* sense of connection to one's environment, regain life balance
- *Interpersonal/social:* communication skills, relating to peers, cooperation, team work
- *Community integration:* appropriate behavior in a retail environment (purchasing the plants), relating to others who use the garden
- *Recreation/leisure:* learn transferable skills to be used at home, interest in a new hobby

Contributed by K. Kennedy

be more appropriate. In long-term care facilities, most individuals will be at the site for an extended period of time. From a horticultural therapy perspective, this provides opportunities for activities that carry over from one week to the next, providing a continuity to the program. Propagating cuttings for dish gardens or terrariums or planting in the spring for a fall harvest are just two examples of long-term activities.

Short-Term or Acute Care

In short-term care, the emphasis may be for a patient to build endurance, increase skills, or to cope with a medical condition or newly acquired illness. The therapist may take a psychoeducational approach, providing training in these and other areas. Ideally, activities need to fit into the time frame of a patient's stay. When incorporating long-range projects such as planting the outdoor garden or forcing bulbs, the therapist would need to help the participant understand they are part of a bigger picture. Although they might not be available for harvest or full bloom, it should be communicated that they are an important part of the process and play a vital role in it. They can also benefit from work done by previous program participants. In these situations, the therapist can remind the patient of the goals they have set and explain how those goals can be met by participating in only a segment of a long-range activity.

Psychiatric

A therapeutic focus using plants as metaphors to promote self-care, life skills, or a plan for developing responsibility (thus building self-esteem) could be key in these programs. Activities need to meet the safety requirements of the facility. This could involve the materials and tools used as well as considerations regarding certain medications.

Physical Rehabilitation

A therapeutic focus of rebuilding motor skills, self-esteem, spatial relationships, cognitive development, stamina, improved concentration, and use of adaptations is used in this setting. The suggestions found in short-term and acute care could also be useful in this setting.

Schools

Study of the environment and group cooperation/education are often the focus in school programs. Activities can be selected that help to meet these goals, at the same time developing and improving social and coping skills. Horticultural therapy programs in high school settings often provide pre-vocational training for students with disabilities. In these programs basic horticultural knowledge and work skills are emphasized.

Corrections

Programs in the field of corrections are typically vocationally based, teaching the job skills required for greenhouse, nursery, or landscape employment. Often large gardens are incorporated into the plan allowing inmates to raise vegetables to be shared with people in need through food banks or homeless shelters.

Wellness or Health Promotion

In the wellness arena, a holistic approach is taken in which horticultural-based activities focus on stress reduction, coping skills, regaining choice and control. Reintegration with nature can result in a regeneration of energy and self care.

Community Gardens

Horticultural therapy can be incorporated into a community garden by using the garden as a site for any type of program. Many community gardens focus on community development and serve disenfranchised populations as well as people with a passion for gardening. Coordinators may develop specific measurable goals and objectives for participants and create a program based upon those. Some areas of focus that would fit well into a community garden setting include (1) activities with an *educational* focus regarding proper nutrition, (2) a *therapeutic role* using metaphors in the garden for life skill development, and (3) a *social/emotional* role in adding garden-related recreational activities to the schedule.

Needs Assessment

Another key factor that influences the selection of horticultural therapy activities or tasks is the individual need or issue to be addressed. In many settings individuals are given some form of a needs assessment when they begin treatment, as described in Chapter 2. This identifies a person's strengths and areas of need often called treatment issues. Further examples of some treatment issues that might come from such an assessment are needs for

- increased socialization,
- increased display of appropriate social skills,
- increased orientation to time and place,
- increased ability to follow multiple-step directions,
- increased healthy sense of responsibility and decision-making skills,
- increased cognitive skills,
- decreased anxiety, and
- increased sensory stimulation.

Most often, a needs assessment will include not only abilities and treatment issues but also information regarding a person's hobbies and gardening interests.

Participants' Backgrounds, Likes, and Dislikes

The horticultural therapist may also provide an assessment designed specifically for the horticultural therapy program. This would help in obtaining a clearer understanding of a person's interest and experience regarding indoor and outdoor gardening. For gardening to have therapeutic value, it must be meaningful to the individual participant. This will occur when the individual's needs are assessed and correctly matched to an appropriate component of horticulture. This assessment also determines if there are potential challenges such as an aversion to getting dirty or allergy issues. It is only natural that someone will participate if they are interested in what is being done. Ideally, a horticultural therapist takes into account the background of those being served, finding out the extent to which they have gardened in the past, the types of projects they enjoy, and those they do not.

When beginning to work with an individual, selecting familiar, enjoyable activities will help to create a sense of safety and belonging for the participant. After a person has been in the horticultural therapy program for a period of time, it might be beneficial to expand the focus of activities to include new and unfamiliar ideas. Providing the opportunity to learn new skills and new concepts can enhance a client's overall functioning level. In the "person-centered care" model (as described in Exhibit 3.4), understanding a participant's background experiences and knowledge is key to providing purpose and meaning to their current lives. This background knowledge is

EXHIBIT 3.4.
Practice Perspective: Person-Centered Care

Person-centered care is a model of care that focuses on the value of each individual. It involves respecting and honoring the uniqueness of each person and allowing him or her to be involved in decisions that impact his or her life. Traditional care and person-centered care differ in the following ways: disease focused versus person focused; managing behaviors versus heavier acceptance; caregiving versus care partnering; control and losses versus empowerment and abilities programming; activities versus meaningful occupation. According to Virginia Bell and David Troxel (1999), the goal of person-centered care is to move the person, even momentarily, from loss to fulfillment, loneliness to connectedness, sadness to cheerfulness, confusion to orientation, worry/anxiety to contentment, frustration to peacefulness, fear to security, paranoia to trust, anger to calm and embarrassment to confidence.

also valuable when attempting to understand what is behind certain behaviors.

Situational Assessment

After the general assessment has been completed, the horticultural therapist may carry out a situational assessment. Formal or informal observations are made of an individual actually performing horticulture activities. This type of assessment can provide an array of useful information about how the person approaches and accomplishes each activity. It can be used to document skills and behaviors seen in horticultural therapy that may be different than in other therapy environments. In the vocational setting (or any task-oriented setting), this may be used to gather data about an individual's work potential in a particular job setting. In this process, the therapist creates a written work sample form that includes the following categories:

- Task
- Objective
- Supplies needed
- Directions to the evaluator
- Directions to the client

- Areas of focus
- Guidelines for scoring

From the data collected, the therapist can then select activities or tasks as part of a training program for the clients that will help them to attain the necessary skills required for the desired job.

Program Continuity for Individual Development

Continuity in the horticultural therapy program is a building block for successful individual development. When a program goes beyond mere activity and becomes a process in which activities or projects interconnect and build upon each other, there is a stronger foundation for a person's growth. Continuity adds that dimension to create an identifiable order rather than just a collection of random events. The medium of horticulture is ideal for providing this connection, as the garden progresses naturally through the seasons. Continuity can be achieved through

- programming on a regular basis, ideally at least once a week;
- selecting activities that build on one another;
- revisiting past projects to note their changes and to orient to time and place;
- using the garden's natural seasonal progression;
- keeping in mind that the activity is part of something larger; and
- maintaining consistent work-space and storage areas.

ADDITIONAL CONSIDERATIONS
FOR ACTIVITY PLANNING

In addition to the aforementioned considerations, planners should also take into account the resources available at each horticultural therapy site and the number and type of staff to be present. Choices that minimize risk to all who come into contact with program spaces and materials are essential. Therapists must be flexible and use evidence-based practice in selecting and presenting activities.

Resources and Restrictions Within a Setting

The setting in which the program takes place also plays a role in activity selection. Following are some of the key factors to take into consideration:

- The time frame allowed for the activity
- Budget
- Other conflicting uses of the programming area
- Availability of water
- Sizes of outdoor and indoor gardening areas
- Hardiness growing zone of the setting
- Materials, tools, and furnishings available
- Number of participants
- Average length of stay at facility for program participants (for example, a typical stay might be two weeks in short-term psychiatric facilities or women's shelters).

Staffing

The number of staff and/or volunteers available to assist with a session helps to determine what activities are selected. Activities that involve moving through the garden should ideally have a staffing ratio of one-to-one for individuals needing mobility assistance. This ensures safety and a sense of inclusion for all those participating. Those activities requiring extensive materials preparation should be saved for times when volunteer help is available. Projects that involve a series of steps or procedures are best used when there is staffing support to distribute materials and provide assistance where needed. Finally, there is the need for plant care between sessions. In programs in which the horticultural therapist is only part time, the level of commitment by staff or volunteers to water and maintain the plants will determine the types of plant activities selected. A strong commitment would allow for activities that require high maintenance, such as propagating seeds indoors or planting a sizeable garden. A minimal commitment would direct the therapist to provide activities that can be taken by the participant at completion of a session, thereby eliminating the need for after-care by staff.

Risk Management

In selecting activities, there are criteria to be considered regarding safety in the horticultural therapy program. It is essential, when working with populations that are confused, very young, or at risk of self-injury to use only nontoxic plants. A list of "unsafe" plants is often available through local poison control centers and Internet sites. A rule of thumb is, "when in doubt, don't grow it." Likewise one should have an awareness of any toxic properties of materials used in activities or in the garden, such as fertilizers or pesticides. The use of sharp utensils and materials is also an area of concern with some populations. Supplies should be used under supervision of staff and tools counted before and after the session has ended.

Another area of concern is photosensitivity (or sensitivity to the sun). There are a number of pharmaceutical drugs with this side effect that can result in serious sunburns. Some individuals may have intolerance to heat and/or cold due to their medical condition. For example, people with multiple sclerosis may have temperature-related issues. Respiratory issues must be taken into consideration when working with fragrant flowers, herbs, or potpourri mixtures. Caution is necessary when using these materials with people with asthma, for example. The horticultural therapy program often incorporates food tasting, so the therapist should be aware of any dietary restrictions and food allergies.

Safety in the garden, greenhouse, and activity area is of utmost importance. Pathways and aisles should be well lit and clear of obstructions so as to be easily traversed by people in wheelchairs or with other mobility issues. The garden and greenhouse areas may also need to be secured in settings such as prisons, dementia care, or some mental health facilities. A careful review of initial assessments and clear communication with nursing staff and the participants themselves are essential in providing a safe and successful experience.

Flexibility

In addition to the planning suggestions given in this chapter, the therapist should be flexible. An effective therapist has an alternative activity planned (Buettner and Martin, 1995).

In the field of horticultural therapy there are many variables. Diversity of participants, weather changes, staffing, and rate of plant growth are just a few of the many factors that can affect plans. Activities can be modified as an effective means of accommodating a group of participants with various ability levels to interact together in the same setting. Adapting the environment, substituting materials, using only parts of activities or adjusting the type or amount of facilitation provided are a few of the key means of adaptation. (More information about adaptations and modifications is found in Chapter 4.) Being able to adapt activities is helpful as well as having materials for one or two alternate activities in storage for those times when a complete change is warranted. Some suggested sessions for those times are games such as garden bingo, grooming and feeding plants, pressing flowers, or garden flower arrangements. These, and sessions that involve direct gardening experiences, take a minimum amount of preparation time.

Evidence-Based Practice

Ensuring a high-quality horticultural therapy program is best done through evidence-based practice. This involves keeping up to date with the latest techniques that have been documented to be effective. Using "tried and true" methods assures the therapist and those being served of a greater likelihood of success. It also proves to be more cost effective through a savings of time and materials that might otherwise be spent on less effective activities. This attention to effective methods adds credibility to the program and to the profession. As more people look to what is being done in the field, more horticultural therapists will start reporting what has been effective in their situations, helping to shape the future of the horticultural therapy profession. One primary source for this evidence-based information is the American Horticultural Therapy Association (AHTA). AHTA produces the *Journal of Therapeutic Horticulture* and numerous information packets as well as an annual conference. Other resources include: regional conferences on horticultural therapy and related areas, colleges and institutes that provide horticultural therapy course work, and colleagues in the field. The following list offers reasons to use evidence-based practice.*

- Ensures quality for consumers
- Keeps one up to date to use the best practices available

---

*Contributed by S. Sieradzki

- Uses tested and true methods
- More cost effective for consumers and third-party payers
- Adds credibility to the HT profession
- Self-pride from knowing you are doing the best possible work
- Will increase reimbursement, which will translate into more HT jobs available
- Ability to eliminate methods that are less effective
- Help shape the future of the HT profession

Session Review

Reviewing each session is just as important as carefully thinking through the setup and process of the session ahead of time. The review process helps the therapist prepare for the next session. This procedure elevates the effectiveness of the overall program and ensures that participants' needs are being addressed in the best possible way. Many variables can affect the outcome of the therapeutic experience. Review by asking questions about each component of the horticultural therapy session, including the *time*, *location*, the *activity process*, and the *therapist*. See Exhibit 3.5 for session review considerations.

The review process is a worthwhile endeavor because of its potential impact on improving the effectiveness of a horticultural therapy session. By answering questions like those in Exhibit 3.5, it is possible to isolate those individual elements that affect the session. With this understanding, the therapist can then more easily begin to change what is not working well. Think about your own setting and create a checklist or review form for yourself, using all of the variables present in your own situation.

PLANNING FOR SEASONALITY AND EFFICIENT USE OF HORTICULTURAL RESOURCES

Once the setting and purpose of the program is established and goals have been identified, in some settings the therapist may create a calendar of activities for the year. This process helps the therapist to keep the activities in alignment with the goals of the program and the

EXHIBIT 3.5. Tips for Practice: Session Review

Begin by looking at the *timing* of the session. Did you have maximum attendance? Sometimes meals, medication, and transportation (internal or external) can affect participants' alertness or their ability to focus on the group session for example. The availability of support staff to help with transportation to and from the session or to provide assistance during the session can affect success just as much as how the therapist utilizes assistance during the session. By just considering when the session is held, a number of variables are affected.

Evaluate the space or *location* of the session in the same way. Look at size and whether the participants were able to maneuver comfortably. Was the seating arranged to facilitate the group process? How were the lighting and the temperature? Was it free of distractions? Consider other questions particular to your organization.

Then take a look at what happened during the session—the *activity process*. Did the intended outcomes occur? Did everyone have an opportunity to participate at the maximum level possible? Maximum participation includes having ample materials and tools, appropriate modifications, and being prepared for the unexpected. It also includes having enough staff or volunteer support to enable individual participants to work on personal goals that might require standby assistance, verbal cues, or visual monitoring for safety. This is also the time to look at timing or flow of the session. Were the participants rushed or was there adequate time for the discussion and activity?

Perhaps the most difficult part of the review process requires the *therapist* to honestly look at his or her own actions, inactions, and words used during the session. For example, consider whether the instructions were presented at an appropriate level and in a logical order. Was the tone of voice used respectful rather than condescending; encouraging of participation rather than stifling? Was the best leadership style used to achieve the desired outcome? What about the group dynamics? Were you aware of and able to manage interactions among the participants? Sometimes knowing what to say to keep the session flowing is difficult. If that is challenging for you, try thinking through the introduction to a topic or the steps of a task ahead of time.

Contributed by K. Kennedy

individuals being served, and to incorporate horticultural resources in a meaningful way. An activity calendar also produces a congruent program that flows with the cycles of nature and helps develop an efficient means of purchasing and preparing supplies.

Creating a Seasonal Calendar

Creating a seasonal calendar is an excellent way to establish order in a horticultural therapy program. Following are considerations in developing this calendar:

- List the major holidays that are relevant to the individuals or community being served. See Appendix B for a list of possibilities.
- List horticultural/nature days of note: Arbor Day, Earth Day, solstices, Horticultural Therapy Week, etc.
- List celebrations or special events that the organization/facility has on its calendar.
- Research miscellaneous celebrations that might fit in with the population being served. Activity therapists and recreation therapists often subscribe to newsletters that share this information.
- Just as each month has a birthstone associated with it, there is also a flower associated with each month. This could be easily incorporated into the program plan.

Creating a Planting Calendar

Once the seasonal calendar is developed, an indoor/outdoor garden-planting schedule can be created. Initially, determine what plants will be needed for specific holiday projects or plant sales; for example: Mother's Day potted plants, holiday amaryllis display, Valentine topiary, bedding plants, or cut flowers. Will plants be needed for miscellaneous plant projects such as terrariums or dish gardens? All of these activities require previous activities that involve propagation of the needed plants. The Mother's Day plants and the Valentine topiary need to be propagated no less than three weeks in advance of their use and the amaryllis need to be forced in late October for a December display. See Table 3.1 for a planting schedule of selected holiday projects.

The outdoor garden schedule should include plants and planting dates for cool and warm seasons. Identifying when the plants should be transplanted to the garden will determine when to start the seeds indoors. Seed packets as well as books on propagating from seed provide information on seeding times. Once these planting dates have been determined, activities can then be developed for the days remaining open on the calendar.

TABLE 3.1. Tips for Practice: Sample Planting Schedule for Holidays

Holiday	Plant Project	Propagation or Preparation Method	Timing (weeks prior to event)
Valentine's Day	Ivy heart topiaries	Tip cuttings	4
St. Patrick's Day	Decorate pots of shamrocks	Plant corms or division	8
Mother's Day	Blooming marigold pots	Seed	8
Father's Day	Succulent dish garden	Division and tip cuttings	4
Fourth of July	Red, white, and blue garden flower arrangements	*Statice*—seed *Celosia*—seed Dusty miller—buy transplants	*Statice*—16 *Celosia*—12 Dusty miller—6
Fall solstice	Fall leaf art	Press	2-3
Halloween	Spider planter	Offsets	2
Thanksgiving	Pumpkin planter (Swedish ivy or other hanging plants)	Tip cuttings	3
Winter holidays	Blooming amaryllis and paper white narcissus	Forcing bulbs	8-10

Coordinating Outdoor with Indoor Horticultural Programming

Ideally program participants should grow most of the items used in the horticultural therapy program. To provide program continuity, the materials used in a winter floral project should be grown and preserved by the participants themselves or in settings where stays are short by those who gardened before them. Throughout the growing season flowers can be pressed or dried, *Helichrysum* (strawflowers), *Gomphrena,* and other everlasting plants can be harvested, wired, or hung. Herbs can also be harvested and dried. Floral projects and crafts provide an opportunity for creative expression as well as an opportunity for gift giving and continued connection with the garden. Prior to a frost is the time to start bringing the garden indoors. Table

3.2 illustrates ways to accomplish this and in turn create a rich program. It is wise to treat all cuttings and potted plants with a soap-and-water solution before bringing indoors to prevent the introduction of insects. Putting the garden "to bed" for the winter is also an important

TABLE 3.2. Tips for Practice: Bringing the Outdoor Garden Indoors

Plant Name	Cuttings	Seeds	Pressing & Drying	Drying & Teas	Potpourri, Oils, Etc.	Cook	Oils & Vinegars
Begonia sp.	X						
Celosia sp.		X	X		X		
Coleus sp.	X						
Herbs	X	X		X	X	X	X
Everlastings—*Helichrysum, Limonium,* etc.		X	X		X		
Impatiens sp.	X						
Lemon Balm—*Melissa officinalis*				X	X	X	X
Marigold—*Tagetes* sp.		X	X		X		
Mint—*Mentha* sp.	X			X	X	X	X
Nasturtium—*Tropaeolum majus*				X		X	
Geranium - *Pelargonium* sp.	X		X				
Rose—*Rosa* sp.			X		X		
Scented Geranium	X		X	X	X	X	X
Wandering Jew—*Tradescantia* sp.	X						
Verbena sp.			X				
Vegetables		X				X	
Zinnia sp.		X	X				

process to share with program participants. Incorporating all of the life cycles of the plants and garden in the program offers individuals a greater understanding and acceptance of life cycles in general (including their own). That theme can be carried on indoors through activities such as artwork, poetry, or garden word games.

SOURCES FOR SESSION IDEAS

An important task for a horticultural therapist is to come up with new ideas for activities. Sources for session ideas abound, including the garden itself as well as the imagination of the program leader. Some books are available that show activity ideas, and many other resources available to the therapist as shown in Appendix B. An important consideration when looking at a potential activity is: "Can the idea be adapted to meet the needs and abilities of the people being served?" (See Chapter 4 for further information about adapting activities.) It is essential to remember that the horticulture activity is the tool to assist people toward their goals. Those that require minimal intervention or support by the therapist or volunteers (that maximize independence of the client) are the most effective. An activity should be age and ability appropriate and should be a source of pride for the client. In a greenhouse or nursery program where items are being produced to sell, the quality of the item is important and projects can be selected that fit the abilities of those doing the work.

Horticulture Ideas

A vast array of gardening magazines and television shows provide the stimulus for plant activities. Other excellent resources are programs that produce materials for youth gardening. The National Gardening Association and the Cooperative Extension Service are two examples of such organizations. Youth gardening information is most often focused on the outdoor garden and is written in a way that is easy to adapt for specific populations. Creating a relationship with managers of greenhouses can be an asset for tours and plant materials from which to develop a project. Criteria for selecting horticultural activities are as follows:

- Activity has therapeutic value and meaning.
- Plants used are safe (nontoxic, no thorns, etc.).
- Plants have the ability to grow in the program environment. Consider plant needs for
  —shade or sun,
  —temperature indoors or outdoors,
  —drought tolerance, and
  —growing space.
- Activity works within budget constraints and time frame. (For example, can the plants be propagated or will they need to be purchased?)
- Plants are relatively easy to grow and maintain.

Floral Design and Horticulture Craft Ideas

Magazines, television programs, and nature craft books are excellent resources for floral design and horticulture craft ideas. Taking time to walk through craft stores, craft shows, floral shops, and garden centers often pays off with new ideas. Networking with activity professionals is beneficial as they have access to many plans for activities, some of which could be adapted to incorporate a horticultural element.

When selecting activities in the craft area be sure the project has therapeutic value and it can be adapted for age and ability appropriateness. Both the cost of materials and the possibility to raise the materials in the garden, or find them in nature, should be considered.

Nature Studies Ideas

Nature or environmental studies can either be programs on their own or enhancements for more diverse programming. Gardens are wonderful arenas for observing and connecting with nature. National organizations such as National Arbor Day Foundation and Earth Day Foundation are just two examples of organizations that provide classroom materials and often plants for use in the horticultural therapy program. The USDA Forest Service is a resource for nature posters, useful in teaching environmental concepts as well as decorating the session area. This is another area in which many materials for use with schoolchildren are available and can be adapted to meet the needs of specific populations. Many communities have nature-related organizations, such as nature centers or public/private parks, that can serve as tour sites or as resources for project ideas and materials. Botanic gardens and arboreta serve the general public with on-site tours, classes, and printed information. The National Audubon Society has excellent information to share on birds and often will supply bird feed to nonprofit agencies. Criteria for selecting nature studies or environmental activity area are: the therapeutic value of the activity, the accessibility of tour sites, and whether the activity is age and ability appropriate for the population served.

SUMMARY

In summary, a well-developed horticultural therapy program focuses on the individuals being served, their therapeutic goals, and their individual likes and dislikes. Further criteria in activity selection include

- therapeutic value,
- the type of setting,
- staffing,
- risk management,
- providing continuity,
- planning for the seasonal calendar,
- adaptability, and
- recorded success.

Following the guidelines outlined in this chapter will help to ensure a safe and successful horticultural therapy program, one in which the participants, gardens, staff, family members and facility all thrive.

REFERENCES

Bell, Virginia and David Troxel (1999). The other face of Alzheimer's Disease. *American Journal of Alzheimer's Disease 14*(1):60-64.

Borg, Barbara and Mary Ann Bruce (1991). *Group system: The therapeutic activity group in occupational therapy.* Thorofare, NJ: SLACK, Inc.

Buettner, Linda and Martin, Shelley (1995). *Therapeutic recreation in the nursing home.* State College, PA: Venture Publishing.

Chapter 4

Working with Program Participants: Techniques for Therapists, Trainers, and Program Facilitators

Karen L. Kennedy
Rebecca L. Haller

INTRODUCTION

This chapter explores techniques that are used to work with participants in horticultural therapy programs. The concepts and methods described are used in many types of health-care and human service fields, and they provide a basic foundation to promote positive outcomes. These approaches are also useful for group leaders, teachers, and program facilitators. By employing them, the therapist/leader will be able to maximize the growth and functioning of those served. Techniques addressed include

- facilitation and group leadership,
- therapeutic use of self,
- motivation and behavior management,
- training methods, and
- adaptation and modification.

FACILITATION AND GROUP LEADERSHIP

Although not exclusively, most often horticultural therapy is practiced in groups varying in size from three or four to fifteen or more in-

Horticultural Therapy Methods
doi:10.1300/5676_04

dividuals. Effective planning, facilitation, and group leadership techniques are therefore key skills for all horticultural therapists, as well as leaders of community horticulture programs. Some considerations for planning and leading groups are the populations served, the personalities involved, the purpose and type of group, the activities or tasks to be performed, and the setting of group sessions.

A collection of individuals can be considered a group when they identify themselves as a group, interact, and have a shared purpose (Austin, 1991). Not all horticultural therapy groups are able to achieve this sense of "being a group" due to factors that interfere with this identification. For example, group members may change frequently, members may only be capable of or allowed minimal interaction, or the group may only meet for one or a few sessions. By consciously cultivating group formation, leaders use a tool that can foster individual growth. Group practice offers a distinct advantage to participants to get their needs met through interaction and relationships with others (Toseland and Rivas, 2001). In horticultural therapy groups, participants develop relationships with the therapist, the plants, and other group members. This group interaction provides opportunities for support, feedback, social role practice, and a wide array of social skills such as cooperation, communication, and trust.

Types of Groups

This section looks first at the types of groups and how they might be used in horticultural therapy programs. Groups can be classified as *activity* or *support groups,* or a combination of the two (Finlay, 1993). Activity groups may focus on tasks or on social experiences. Task activity groups focus on the horticultural activity or task to be performed. Goals may include development of attention span, concentration, completion, cooperation, creativity or vocational skills. The end product of the group effort is valued and motivating. In social activity groups, the emphasis is on leisure interests and social interaction. Participants are encouraged to interact, cooperate, and have fun. Activity groups are perhaps the most prevalent type in horticultural therapy practice. Support groups, on the other hand, focus on communication and/or psychotherapy. Communication support groups aim to share experiences and feelings in order to provide support for group members. Psychotherapy support groups focus on reactions, interpersonal skills, and feelings. They may help members address psychological issues and work on personal goals within the group. Examples of these four group types as used in horticultural therapy are summarized in the following list:

- Task activity group—group with gardening or vocational production focus
- Social activity group—garden club at assisted living facility
- Communication support group—group that uses gardening metaphors and shares experiences
- Psychotherapy support group—group gardening used to explore individual feelings

In practice, horticultural therapy groups may include one or more of these foci. For example, in a "garden club" at a long-term care facility, residents may learn gardening skills, socialize and have fun, talk to each other in supportive ways, and perhaps share feelings as part of the session. In this setting and others, the emphasis of the group may shift depending upon the current needs of the people involved. In effect, the group then functions as a task and social activity group as well as a communication support group.

Leadership Styles

In order for horticultural therapy groups to be successful, the horticultural therapist must lead the group effectively. The leader's planning, preparation and style all impact the achievements of the group. The objectives are to create and guide the group in ways that maximize growth and development of participants. Leadership styles necessarily vary according to the purpose or type of group, the level of group functioning, and the horticultural activity performed. Various situations call for a diverse repertoire of roles to be adopted. The leader/therapist may act as a teacher, planner, coach, facilitator, or motivator at any given time. For instance, during a support group for people coping with cancer, the therapist assumes the role of a teacher when providing cultural information about tropical plants and techniques for designing and planting a dish garden. As the clients discuss how their choices are analogous to life choices they've made and balance in their own life, the therapist's role shifts to coach or facilitator of the discussion. Thoughtful consideration and flexibility in adapting to the present need is required.

Three main styles of leadership are commonly described—*authoritarian,* democratic, and *laissez-faire* (Austin, 1991; Finlay, 1993). They represent a continuum of directive to nondirective approaches, each with distinctly different results. The authoritarian style is characterized by strong control over the group's actions, with each step directed and closely supervised. At the other end of the continuum, using the laissez-faire style involves very little direction by the leader, with group members deciding and initiating action on their own. A democratic leader outlines the task at hand, provides some guidance, and encourages the group to discuss how and what to do. While therapists may be most at ease using one of these leadership styles, it is important to become comfortable with the full range of styles in order to facilitate growth and group functioning in the many situations encountered in horticultural therapy programs. Each style offers advantages and may be the appropriate choice for the situations described in Table 4.1.

In choosing a leadership style for each situation, the horticultural therapist should consider the functioning level of individuals and the group, the setting, the type of horticultural task to be performed and the relative importance of the end product. For example, an authoritarian style may be appropriate in a vocational setting with strict quality standards or time constraints or in a setting where safety issues are paramount. At the other end of the spectrum, a small horticultural therapy support group for people with cancer may benefit from a laissez-faire style of leadership with an emphasis on creativity and individual choice.

Fostering Functional Group Behaviors

Group members perform two major functions as described by Austin (1991)—*task functions* and *social-emotive functions.* When used as a description of group dynamics, task function refers to those activities that help members to achieve their goals, whether these are cognitive, physical, emotional, social, or vocational. The social-emotive function of a group is to promote and achieve a positive group atmosphere. In order for groups to perform both of these functions well, the horticultural therapist must provide effective settings,

TABLE 4.1. Tips for Practice: Leadership Styles

Style	Description	Useful style when...
Authoritarian	• Directive approach • May foster dependence • Responsibility is with the leader	• The group is newly formed. • There is a limited time to accomplish task. • High standards exist for task performance. • The group is very large. • Participants have limited social skills. • Participants have cognitive deficits. • Negative/disruptive behaviors occur. • Safety is paramount. • Structure is essential for group function. • Choices are threatening or overwhelming to clients.
Democratic	• Members involved in decision making • Need time for discussion • Feelings of teamwork	• Time allows for discussion to take place. • Social skills are key goals of the group. • The group is functioning well. • It is important to seek full participation.
Laissez-faire	• Nondirective approach • Open and permissive • Client centered • Group members initiate action • Emphasizes independence	• Creativity is emphasized. • Trust and responsibility are goals. • Group members are able to give and accept social influences. • The group size is small. • Group members need to set their own agenda. • Standards for the end product are flexible.

structure, and leadership, and also manage any nonfunctional behaviors that occur. Horticultural therapists face many kinds of challenges in leading groups, including those associated with participation levels and disruptive behaviors. Exhibit 4.1 provides some questions to ask when participation is limited or interfering.

EXHIBIT 4.1. Tips for Practice: Participation Levels

Participation levels can range from nonattendance or lack of participation to dominating discussion and attention. Questions to ask in these situations include the following.

- Do the facility staff and co-workers support attendance through transportation, scheduling, reminders, etc.?
- What factors motivate attendance and participation? For example, do the time of day, location, and use of end products contribute to full participation?
- Are horticultural therapy sessions voluntary or mandatory? Are the sessions tied to a reward system for clients?
- Do medications affect alertness?
- Are there other appropriate outlets for personal discussion, counseling, or problem solving?
- Are quiet members encouraged to be involved?
- Do domineering members receive sensitive feedback regarding their behaviors?
- Has the client described a barrier that might prevent participation?
- Do financial concerns prevent attendance?
- Are activities appropriate and interesting?
- Are group members compatible?

Disruptive behaviors may include interruptions, displays of anger or aggression, or bizarre or socially inappropriate actions. For example, interruptions could consist of continually asking for help, shouting, walking away, or interrupting another's statements or conversations. Displays of anger or aggression could include shouts, verbal or physical threats, destruction of property, or crying. Socially inappropriate behaviors may involve approaching strangers with hugs, psychotic or delusional episodes, or overreactions to common situations.

In these and other disruptive situations, the role of the group leader or horticultural therapist is first to manage the behavior sufficiently to reduce the real or perceived threat to the perpetrator and the other group members. The group should be removed if there is immediate physical threat. In general, redirecting the individual to appropriate action is effective. A person with dementia who is experiencing agitation may respond to the therapist's calm voice and a simple gardening task in which to be engaged. By recognizing signals that indicate that a group member's anger is escalating, the leader can take action before a crisis occurs. A physically active garden task, such as digging, could be assigned. Conversation, appropriate expression of feelings, relaxation exercises or other techniques could be used. Harvesting leaves or flowers from a scented plant such as lavender, whose fragrance is known to be calming, or a repetitive task such as weeding or cultivating the soil are examples of garden-specific tasks for eliciting behavior change. By knowing the participants, the therapist may also select a task the individual particularly enjoys, such as watering. The garden or greenhouse environment offers many options for calming and redirecting disruptive behaviors.

THE THERAPEUTIC USE OF SELF

The primary components of a horticultural therapy program are the plants and growing environment, client, and the therapist. The development of client goals and the plant-related activity selected to facilitate achieving the goals are the foundation of the program. The role of the therapist in this process cannot be underestimated (Schwebel, 1993). In fact, the relationship established between the therapist and the client is a crucial element in the overall process.

Therapist's Role

The therapeutic relationship differs from a social relationship in that meeting the needs of the client is the basis of the relationship. To be effective, the therapist must be aware of his or her own attitudes, feelings, biases, values, and beliefs and how these affect interactions with each client. The purpose of the relationship is to address the client's goals. To facilitate this process, the therapist must

- use effective communication techniques,
- encourage a satisfying relationship,
- provide physical and emotional support,
- facilitate the client's understanding about himself or herself, and
- understand the therapeutic process.

Other elements of a therapeutic relationship are

- rapport,
- respect,
- empathy,
- genuineness,
- authenticity,
- trust, and
- patience.

The Use of Self

In creating a nonjudgmental atmosphere in which optimal client growth can occur, rapport is built on *warmth* and *acceptance*. Verbal and nonverbal communication reveals the therapist's feelings and attitude about clients. This includes facial expressions, posture and tone of voice. Clients connect with therapists who present themselves authentically and who are honest, open, and real with them. Regardless of personal feelings about a client's choices, behavior, personality, or appearance, treating the client with respect means that the therapist believes unconditionally in the positive value of that individual.

Self-disclosure is sometimes useful and appropriate in establishing a therapeutic relationship. When exploring feelings or experiences in common, it may be helpful in expressing empathy. However, sharing personal experiences should be done carefully, not as a venting opportunity, so as not to lose focus on the client's needs.

Creating boundaries in therapeutic relationships establishes trust and safety. Be mindful of the use of touch. Not all individuals interpret touch in the same way. Be conscious of body position and personal space. Although some worksites are extremely casual in the horticultural therapy field, a therapist's attire should be benign and professional so as not to distract or interfere in a professional relationship.

Since change occurs slowly and in small steps, exhibiting *patience* is a great form of encouragement and support. This does not mean one should abandon great expectations. Tolerance of the journey taken to achieve the client's goals can communicate support for the individual's efforts in the process and strengthen the relationship.

Finally, an important aspect in the therapeutic use of self to consider is how the therapist's role in the relationship impacts himself or herself. See Exhibit 4.2 for more information on maintaining a healthy perspective.

EXHIBIT 4.2. Tips for Practice:
Techniques in the Therapeutic Use of Self

Ground

It is important to ground oneself when working with others. Without a sense of "groundedness," the therapist might need the client to achieve a specific goal or act in a certain way in order to feel solid within one's self. The horticultural therapist can visualize having a root that extends into the earth. This root can bring balance and ground emotions or concerns that come up for the therapist while they are working.

Focus

It is important for the horticultural therapist to see the client's work within the context of his or her whole life. It is also important for the therapist to see where the client embodies wholeness despite his or her disability, disease, or disadvantages in life. If the therapist softens his or her focus, as if using peripheral vision when viewing the client, he or she will be better able to hold onto the big picture of this person's life. In doing this the therapsit will be better able to assist the client in accessing this experience as well.

Self-Care

The horticultural therapist's ability to help others is connected to his or her ability to take care of himself or herself. It is important for the therapist to have practices and routines that support letting go of the work and replenishment.

Boundaries

It is important to recognize that the horticultural therapist does not heal the client. The therapist facilitates a relationship between the client and plants that supports the capacity for healing.

Blending

The horticultural therapist's task is to help the client identify goals that fit the client's life circumstance. This therapeutic contract requires blending with where the client is now. The therapist does not predetermine what is best for the client. The only time that controlling the direc-

(continued)

(continued)

tion of the work is called for is if the client is in danger of hurting himself or herself or another person.

Closure

Often the horticultural therapist works with a client for only a short period of time and does not see the results of his or her work. It is important for the therapist to create a place for closure with those he or she works with. The horticultural therapist can visualize completing a contract with a person. This supports the recognition that the client's life will continue to unfold.

Contributed by J. S. Rice

Communication Techniques

Attention to the elements of a therapeutic relationship mentioned previously establishes the foundation for effective communication. Helping clients to become more at ease with plants if they fear they have no "green thumb" is often another helpful first step. Pictures, demonstrations, and the plants themselves facilitate the process for most people—especially if memory, abstract thinking, or unfamiliarity is an issue.

During discussion, ask open-ended questions to illicit a more lengthy response and encourage descriptions of thoughts or feelings. Why and how questions tend to put people on the defense. When appropriate, draw on examples from the plant world to illustrate points, make abstract concepts more concrete, or to provide cues as in Exhibit 4.3.

Humor is often useful to create a positive, relaxed atmosphere particularly with clients the therapist knows well. An example of the use of humor is shown in Exhibit 4.4. It is a good technique to respectfully incorporate when a task may be frustrating, to alleviate anxiety or nervousness when trying something new, or when teaching stress-reduction coping skills. Humor is not particularly constructive with people who have difficulty thinking abstractly or who are coping with major depression.

EXHIBIT 4.3.
Examples from Practice: An Oak Tree Cue

A client with a traumatic brain injury who had difficulty with awareness of his body posture used an oak tree as a cue. This client over time would slowly begin leaning so far over he would eventually fall out of his chair. An acceptable, age-appropriate cue to correct his posture was to simply say "like an oak tree." He selected an oak tree because to him it symbolized strength and uprightness. It was also a more dignified cue than hearing the therapist say "Sit up straight" frequently.

EXHIBIT 4.4.
Examples from Practice: Plants and Humor

Humor can be interjected into HT sessions through words or the planting project itself. Consider purchasing or creating containers with faces and selecting a variety of plants to grow as hair, add a nose and glasses to a plant or pot. Make up your own common names for succulents. Make seed packet cards with original puns such as: beet seeds—"my heart beets for you"; lettuce seeds—"lettuce be friends."

People respond to plants in fantastic ways. They are encouraged to try harder, distracted from limitations, motivated by previous (as well as the anticipation of new) experiences, and can achieve myriad physical, psychosocial, and vocational goals. It is the relationship with the therapist that connects the optimal horticultural experience with the client's best interest.

In summary, the primary goals of the therapeutic relationship are to

- facilitate communication,
- facilitate acquiring new skills,
- encourage and support efforts made toward established goals, and
- promote independence.

MOTIVATION AND BEHAVIOR-MANAGEMENT TECHNIQUES

Horticulture is a powerful therapeutic tool for individuals who have interest in and enjoy the beauty of plants and flowers. People respond to the sensory stimulation inherently involved with interacting with plants and flowers (Haas and McCartney, 1996). For others the usefulness of plants for food, crafts, decorating, etc., has appeal. Horticultural therapists use the people/plant response to motivate clients to participate in both group and individual sessions. Horticultural therapy sessions are an ideal venue for positively impacting behavior because of the relaxed and nondefensive physical and mental state participants tend to exhibit during horticultural therapy activities. In rehabilitation settings, for example, physical and occupational therapists, rehabilitation nurses, as well as professionals working on spiritual, behavioral, and chemical dependency issues find their work to be positively impacted by the participant response to the horticultural therapy program (Murrey, Wedel, and Dirks, 2001). When clients are less agitated, defensive, and tense they are more receptive to challenging therapeutic interventions.

Positive Reinforcement

By their very nature, plants respond to care. The direct causes and effects of watering, fertilizing, pruning, and grooming result in healthier plants. New blossoms and fruits are tangible rewards that participants can identify and feel good about. Therapists use this concept to reinforce desired behaviors such as increased

- frequency and level of participation,
- standing tolerance,
- tolerance of task,
- positive interactions with others in the group, and
- refocus on health and wellness.

The horticultural therapist uses positive reinforcement to reward behaviors that are occurring and to encourage desired behaviors. In groups, the therapist can manage behavior and encourage pro-social behavior by giving group participants tasks they enjoy as positive reinforcement. For example, a participant in a pre-vocational setting has trouble working without arguing with others. If he follows through with weeding for fifteen minutes without arguing with another participant, he will be given the task of watering at the end of the session, which he particularly enjoys. An important part of the process is bringing the participant's attention to the desired behavior and the reward.

The growth participants see within the garden or growing space is used to reinforce progress on goals and desired outcomes. This awareness of growth is an effective strategy to build confidence and self-esteem. Participants may even exceed their goals and surprise themselves at what they can achieve. The process continues to build as participants accept new responsibilities and advance their progress on cognitive, behavioral, and mental health goals. It is useful for horticultural therapists to verbally recognize growth and change of the participant or simply reinforce progress with verbal praise such as "good job."

Horticulture As a Motivator

Growing plants is an experience participants are likely to be familiar with before they are part of a therapy program. It is viewed as a meaningful activity, especially to those who have lost their independence as a result of an injury, illness, or living situation. Taking plants back to their living space provides participants with a sense of control over their environment. Participants may find their own meaning and motivation in the planting process or activity by drawing comparisons or analogies to their own life situations. Pruning the plant to encourage new growth and free it from tangled overgrowth illustrates the pruning one can do in life to get free from any parts that are overwhelming.

In some programs, being allowed to participate in the horticultural therapy program is the reward for good behavior. In correctional facilities for example, inmates must demonstrate nonviolent behavior and adherence to the rules before being admitted into the program. This behavior must continue as they learn new vocational and coping skills that will enable them to be successful citizens when released. Programs that facilitate behavior change also help clients identify obstructions to a healthier and better life. With the support of the horticultural therapy group, clients learn how to cope with these barriers and begin to weed them out.

The basic activity of plant care may be used to provide reality-based situations to motivate self-improvement. For example, if due to poor behavior a client is confined to his living quarters and is not allowed to attend horticultural therapy sessions, rather than "rescuing" the plants the therapist may allow the plants to be neglected. This provides an opportunity to learn cause and effect through natural consequences.

In programs in which participants grow and sell plants, the customers provide the positive reinforcement. Participants are often able to share valuable information about plant care to their customers, who may be peers, staff, or community members. This knowledge and the ability to produce something that others esteem and are willing to pay for is empowering and motivating. Participants shift from their role of care-receiver to caregiver by becoming the one providing both information and a valued product to others.

Tips and Techniques

Selecting the most appropriate plants, tasks, and activities is key to maximizing the desired behavioral responses. It is important to match the plants and activities to the participants' preferences and interests. A short interview or survey can yield favorite flowers, colors, or fragrances as well as special interests such as cooking or specialty plants and gardening history. For some participants, familiar plants, flowers, and fragrances are comforting, reinforce self-confidence in new and unfamiliar situations, and trigger reminiscence. New experiences and intriguing plants captivate others. Understanding a client's attitude toward plants and gardening as well as how he or she feels about the living environment or treatment programs is necessary in choosing the best approach for the individual.

Be mindful of gender and ethnic backgrounds. For example, after a presentation on herbs in a mixed-gender group, it may be more effective to plant an herb container garden rather than make fragrant bouquets (such as tussie-mussies or nosegays). Women tend to be more comfortable making bouquets of flowers. The activity of planting an herb container garden is more gender neutral and supports the planned discussion.

Special considerations may exist that are unique to facilities or the conditions of the people served by the program. For example, some people may be sensitive to fragrance due to pulmonary issues or chemotherapy, or some may tolerate natural plant fragrance better than synthetic smells. Others may lack the sense of smell altogether due to a brain injury.

Fortunately, the diversity of the horticultural world offers a plethora of plants and associated tasks that appeal to a wide range of interests and abilities. As described in the section titled "Adaptation and Modification," activities and tasks can be graded to increase in difficulty or complexity as goals are achieved. The safe environment established by the greenhouse or garden may be ideal for learning new techniques. For instance, participants may be more motivated to learn compensatory strategies for memory or cognitive deficits in the safety of the greenhouse or garden setting. Once comfortable with the strategy, it can be transferred to other vocational or living situations.

TRAINING

In all types of horticultural therapy programs the horticultural therapist is called upon to provide teaching or training to individuals and groups. The training may involve specific horticulture skills of various complexity, from planting seeds to pruning fruit trees, or may be a simple instruction such as "find a node on the coleus plant." Or the aim may be to help an individual learn a social skill, a self-help behavior, or other activity. This section focuses on those teaching/training skills that are used to lead individuals and groups in the performance of horticulture tasks or activities. The therapist's role as a trainer or teacher requires an in-depth understanding of the task or activity to be performed, the type of learning desired, as well as specific techniques to support or elicit competent performance.

Task Analysis

Understanding the task or activity involves a process called task analysis. (The terms *task* and *activity* are used interchangeably in this chapter to refer to the horticulture action to be performed by the client.)

The process involves looking at (1) the steps to successfully perform the activity (Lamport, Coffey, and Hersch, 2001), (2) the materials, equipment, and facilities required, and (3) possible adaptation or modification to materials, instruction, or environment. It may also include (1) analyzing the skills necessary to perform the task, (2) looking at the cultural, age, and developmental appropriateness of the task, and (3) identifying therapeutic qualities of the task. This section focuses on the necessary steps to do the task, including developing an awareness of what is entailed and the concrete actions required.

The analysis begins with performance of the task. More experienced gardeners often overlook this essential step. While familiarity with the task or activity at hand increases confidence and contributes to horticultural success, leaders also need to be aware of the meaning, effects, and demands of each activity from the client's perspective. Furthermore, it is important to identify a preferred method for accomplishing the task, as there may be several acceptable approaches to any particular gardening task.

First, as the task is performed, the therapist should notice and record the following: (1) any thoughts or feelings evoked while doing the task; (2) the physical and cognitive requirements to complete it; and (3) any emotional response upon completion.

This helps the therapist/leader to understand some inherent qualities of an activity and possible meaning it may have to program participants. Understand that background, culture, and health status will all affect an individual client's reaction to and experience of each activity.

Second, record the materials, equipment, and space used in the activity.

Last, identify the actions necessary to perform the activity. Write out the sequential steps that were completed in concrete observable terminology. In other words, describe what was done that could be directly observed by another. Include as much detail as necessary to clearly show what took place. It is helpful to watch someone else perform the same task and use the observations to modify the list of sequential steps. See Exhibit 4.5 for an example of a basic task analysis that lists the materials and steps necessary for performing a horticultural task.

The basic performance steps may be modified to allow a client to successfully complete the activity. Changes in the setting, tool, and materials selection, number of steps to perform, degree of self-

---

EXHIBIT 4.5.
Examples from Practice: Task Analysis

Task: Propagation of indoor foliage plants by stem tip cuttings

Materials: Stock plant, pruners, container(s) filled with potting soil, water

Steps:

1. Find direction of growth on stock plant and locate the tip.
2. Measure from the tip of stem with an index finger.
3. With the pruners, cut the stem at the point located about one finger length from the tip. Be sure the cutting point is located below at least the third node.
4. Remove leaves from lower one or two nodes.
5. Push finger into center of potting soil in container to make a hole about two inches deep.
6. Place cutting in hole, with bottom nodes (those that are leafless) below soil line.
7. Gently firm soil around cutting to hold it upright and make soil contact with the nodes.
8. Repeat until desired number of cuttings is completed.
9. Water.

---

initiation required, or time allowed may be necessary for success. Based on knowledge of the client's functional level, the therapist should aim to encourage the most independent performance possible and should gradually reduce the modifications or supports as the client progresses. Modifying the basic task analysis results in an individualized approach for training and performance. (Further information on adaptation and modification is provided later in this chapter.)

The clear outline of performance steps, modifications, and supports makes it possible to have consistent expectations of the client. It even allows multiple trainers to maintain this consistency. In addition, the task analysis may be used to document performance on each step, create a record of progress, and identify the need for further training or adaptation. See Exhibit 4.6 for another example of how the sequential list of steps might be used.

EXHIBIT 4.6.
Examples from Practice:
The Importance of Following Directions

Individuals with traumatic or acquired brain injuries in a vocational training program may have following multistep directions as a goal. In this case it is important for the therapist to determine the best method for accomplishing the task, give clear and succinct instructions, and provide any compensatory strategies necessary for accomplishing the task. A checklist of the steps is an example of a compensatory strategy for short-term memory loss.

For example, Adam may decide that rather than filling a pot, selecting a rooted cutting, transferring it to the pot, and adding remaining soil as instructed he would prefer to do the task assembly-line style (filling many pots at a time, etc). While his method would still achieve the desired end result, he would not meet his goal of following directions. In this case it would be up to the therapist who is filling the role of work supervisor to decide how to manage this individual. Requiring the client to check with the supervisor before making procedural changes reminds the client that following directions is important to succeed in getting and keeping a job. It also addresses communication skills and appropriate work behaviors.

Note: For other types of groups, following directions exactly as given may not be as important as comprehension and the client's satisfaction with final product.

Types of Learning

In addition to analyzing the task or activity to be performed, it is helpful to be clear about what type of learning is required of the client. Is the aim to learn facts, concepts, principles, procedures, interpersonal skills, or new attitudes? Table 4.2 shows teaching strategies for each of these types of content (Kemp, Morrison, and Ross, 1998). To determine the types of learning and the techniques to use, the therapist also considers the therapeutic goals as well as the abilities of participants.

Goals and Objectives

Types of learning and training techniques used should correspond with the objectives of the individuals served. For example, in a voca-

TABLE 4.2. Tips for Practice: Teaching Strategies for Various Content or Performance

Content	Teaching Strategy	Horticultural Example
Fact	Show, practice, and rehearse	Ripe 'Early Girl' tomatoes are red.
Concept	Show, describe, and organize	Identify tomatoes ready for harvest.
Principles	State principle or rule	Plants need healthy roots to grow.
Procedure	Demonstrate, organize, elaborate, practice	Transplant seedlings to pots.
Interpersonal	Model, imagine, rehearse	Cooperatively mix soil and fill flats.
Attitude	Model, imagine, rehearse	Gardening helps me to reduce stress.

tional horticultural therapy program the client may be called upon to learn any of the content types shown in Table 4.2. More specifically, he or she may learn the procedure to transplant seedlings or to work cooperatively on a task—two very different types of learning. The focus depends upon the person's current functioning and treatment objectives.

Skills, Abilities, Experiences of Individuals

The type of learning to be expected is also based on the level of experience of those involved. For example, facts generally need to be understood prior to learning the principles associated with them. Similarly, it is useful for a client to know facts (such as identification of plant parts) before teaching him or her a procedure (taking stem cuttings of houseplants).

Strategies for Teaching Procedures

The basis of horticultural therapy is active participation in a horticulture activity or task. The client performs basic or complex procedures through involvement in the therapy session. Therefore, horticultural therapists frequently are called upon to teach procedures. As seen in Table 4.2, to teach procedures most often the therapist begins with a demonstration or modeling of some sort. A description of this common training technique, as well as verbal and physical methods, is shown in Exhibit 4.7 in order to give the therapist options for meeting the training needs of various participants (Callahan and Garner, 1997).

Depending upon the setting and population served, horticultural therapists may find each of these techniques to be useful. The techniques may also be combined, as when verbal instructions are given along with a demonstration or physical cue. In general, the therapist should choose the technique or assistance that is the least restrictive or interfering in order to be respectful and to encourage independence. Other factors to consider when choosing a training technique include individual learning styles of clients, safety issues, and the relative importance of quality plant production or garden appearance.

Learning Styles

Leaders may accommodate various learning styles by the use of an assortment of teaching techniques. Individuals tend to prefer or focus on one mean of gathering and processing information. They may prefer visual, auditory, or kinesthetic processes to learn and retain information. Similarly, people process information in various ways—learning best by considering broad concepts or in sequential fashion, by intuition or observation, etc.

Safety Issues

Choose training techniques that maintain a safe working environment. For example, physical assistance is necessary to teach a four-year-old child who is blind to cut houseplants for propagation.

Importance of Horticultural Success

Choose training techniques based on production demands and needs for attractive, well-kept garden environments. In other words, weigh the relative importance of the activity process and the end product in the type of horticultural therapy program offered. Keep in mind that a well-kept garden and successful plant growth is generally desirable and motivating to participants.

EXHIBIT 4.7.
Tips for Practice: Training Techniques

Demonstration—shows the object or performance

- Uses a common and natural way of teaching or training
- Shows acceptable methods for doing a procedure
- Requires that the client observe carefully
- Helps learning and self-confidence to see the end result of the product (activity)
- Is often accompanied by verbal information to explain details or context
- Initiates the activity—usually occurs before the client begins

Verbal—uses spoken instructions, directions, or reminders

- Allows the therapist to explain details or bigger picture
- May cue or prompt the client to start the next step or to correct performance
- May create an overload of information, particularly in noisy settings or when mixed with conversation

Physical cue—uses a physical motion to teach or prompt

- Includes:
  —Gesturing—pointing to object or area to prompt next step
  —Modeling—miming the desired motion without words; simulating the necessary movements
- Effective in noisy environments or when language barriers exist

Physical assist—using hand-over-hand motions to assist the client in performing the task

- May be useful with people with severe cognitive, sensory, or physical disabilities
- May be useful when safety is an issue
- Only use if necessary, as this method allows for less independence of the client
- Be gentle and respectful. Do not force. Must gauge reactions of the client.

ADAPTATION AND MODIFICATION

To meet the varied needs, goals, and abilities of the people served in horticultural therapy programs, leaders adapt methods and modify the activities to be performed. As stated previously, the activities planned must be accessible to the participants and appropriate for their ages and abilities. The use of horticulture allows for many gradations and methods to accomplish any given task. Adjustments to an activity are planned for in advance, based on knowledge of each client, and may be indicated in a task analysis. However, in practice, the therapist typically is also required to modify aspects of the task on the spot in response to the individual functioning seen during the activity. For instance, a person may need more instruction than usual due to medication or illness. Someone who just experienced an emotional event might be overwhelmed by distractions or noise in the greenhouse activity space. Also, since most horticultural therapy sessions take place in groups, variations of mood, cognitive functioning, physical abilities, and social skills are regularly encountered. Flexibility and attentiveness to the people served are essential talents of horticultural therapists.

Adaptation

Adaptation involves changing the means to accomplish the task in order to enable the client to be successful. It includes adaptations to tools, the environment, the client's position, and the instruction provided (Hagedorn, 2000). The outcome or end result of the activity is not changed (Lamport, Coffey, and Hersch, 2001).

For example, in the task analysis shown in Exhibit 4.5 earlier in this chapter, the end result of the task itself is that a cutting is correctly planted and watered. The use of round-tipped scissors as a *tool* adaptation may enable a young child to be successful and safe. For someone with autism, changing the *environment* so that the entire sequence of steps is performed individually in a quiet section of the greenhouse or activity room may be necessary in order to enable the individual to focus on the task at hand. (Although beyond the scope of this book, garden design is a type of modification that is also often necessary for horticultural therapy programming.) Performing the task in a standing position, such as at a potting bench, is an adaptation of *position* that provides an opportunity to work on an objective to increase standing endurance. The type and frequency of *instruction* given can also be adapted to the needs, abilities, and goals of program participants. Single-step verbal directions may be necessary for a client who is unable to remember or follow a sequence of steps. Exhibit 4.7 describes other types of instructions used for training that may be considered. In these examples, the purpose of adaptation is to alter how the task is to be accomplished by the individual client—matching methods with abilities.

Modification

In addition to altering the methods used to accomplish a task, the task itself can be modified or graded to increase or decrease the performance demands placed upon the client (Hagedorn, 2000). Incremental modifications may be made as the client progresses or declines, facilitating full participation and goal achievement. Possible modifications or grading to the task of taking stem tip cuttings (Exhibit 4.5) are to

- change the *length of time* allowed for the activity or the number of cuttings expected in a given time period;
- modify the *difficulty* or complexity by preparing the cuttings for potting before the session; or
- adjust the *demands* of the activity (could be physical, sensory, social, perceptual, or cognitive) by altering the social aspects of the task—requiring group members to share tools and materials.

Again, the idea is to modify the task itself to fit the needs and abilities of the person performing the task.

Recommendations

Additional principles to consider when choosing adaptations and modifications to horticultural activities are *normalization* and *empowerment*. The activity should be offered in a manner that is normal and real to the participants. For example, whenever possible hold gardening sessions in the garden itself. Be sure that activities are age appropriate and fit within a normal routine for those served. Clients can be empowered to perform independently by offering choices, enabling garden features, appropriate challenges, and opportunities to participate in planning efforts. See Exhibit 4.8 for examples of how to promote independence and empowerment.

Therapists should use care to modify activities only as necessary, allowing clients to experience the essence of the gardening experience with minimal intrusion (Buettner and Martin, 1995). Modifications should also help individuals and groups to meet stated goals and objectives. Exhibit 4.9 and Table 4.3 show how a basic activity might be changed for specific populations and purposes. Ideas for how to work with people with mental illness are presented in Appendix D. These approaches can be applied in many different settings beyond the mental health arena. For example, under the category of major depression an issue identified is low self-esteem, a concern that may be present across many populations and settings. Note that each idea is intended to address an issue faced by the client.

EXHIBIT 4.8.
Tips for Practice: Promoting Independence

Promoting the highest deree of independence possible is a goal of most horticultural therapy programs. Techniques vary depending on client group and type of program but the results empower, motivate, and improve self-confidence and self esteem. Some examples are:

- Provide choice within an appropriate scope that can provide some control over their living environment and improve their satisfaction over their living arrangements. This might mean providing a choice of three ribbon colors or letting them decide which tomato varieties to plant.
- Supply the materials necessary for the project, but rather than having all of the materials in front of each person, encourage participants to scan the room for items they need, initiate a request for an item, or get it themselves.
- Encourage problem-solving skills by allowing group members to make choices that affect how they accomplish a task, items to add to the garden, or ways to address situations (e.g., aphids on the basil plants)
- Have garden containers or beds at different heights to provide gardening opportunities for those who need to stand, sit in a chair or wheelchair, or sit on the ground.

EXHIBIT 4.9.
Examples from Practice:
Adapting and Modifying Activities

To make a *table-top topiary,* clients create their own frame by ma-nipulating twelve-gauge, vinyl-coated copper wire into shapes. The frame, soil, and ivy plants are placed in an appropriately sized pot. The project is completed when the ivy plants are gently twisted around the frame. Notice how the basic activity is modified and/or adapted for the following groups:

Older Adults in a Nursing Home

The session goals are to stimulate social interaction between the residents and provide opportunities for reminiscing and creative ex-pression.

Residents are directed to create shapes for use during the winter holidays, such as evergreen trees, wreaths, stars, etc. In addition to long ivy plants, residents are provided with ribbon and dried flowers for additional decoration. The discussion includes holiday decorating and traditions. For individuals with arthritis or weak grip, use an easier to bend wire such as a fourteen- to sixteen-gauge wire.

Adults with Chronic Illness

The focus of this support group is to facilitate a discussion that helps participants identify areas in their lives that need support and discuss strategies for shaping their lives in new directions. Clients are directed to create shapes that symbolize the growth and change they intend to work toward. Provide wire cutters and accessories such as dried flow-ers, pipe cleaners, craft sticks, ribbon, etc., to enable maximum cre-ativity.

Adults with Mental Retardation

In a vocational program, adults will be creating topiary wreaths to sell during a spring sale. Clients work on following directions and consis-tency for a saleable product. A template or form will be used to create frame uniformity. A pictorial checklist taped to the work surface will help clients create the frames and plant the topiaries.

TABLE 4.3. Examples from Practice: Adaptations

Disability/Illness	Adaptations
MR/DD	• Limit the number of plant choices. • Use sturdy rather than fragile plants.
Hip or knee replacement	• Raise the planting surface to encourage standing. • Provide a variety of plants and accessories to choose from to encourage attention to the process.
Stroke, left side affected	• Place some materials on the participants left side to encourage visual scanning of the workspace and use of left hand. • Provide several choices to encourage planning and organization.
Chronic Illness	• Provide a wide variety of plant choices to enable participants to select plants that represent their current short-term and long-term goals for health management.

Note: Using the activity of planting a dish garden, this example illustrates how the same activity can be adapted based on treatment needs or to enable independence.

SUMMARY

In order to maximize the potential goal attainment of program participants, horticultural therapists need to acquire and practice an array of skills and techniques to employ in day-to-day interactions and programming. This chapter illustrated

- facilitation techniques,
- therapeutic use of self,
- motivation and behavior management,
- training methods, and
- adaptation and modification.

All are vital tools used by horticultural therapists for effective treatment and positive outcomes.

REFERENCES

Austin, David R. (1991). *Therapeutic recreation: Processes and techniques.* Second edition. Champaign, IL: Sagamore Publishing.

Buettner, Linda and Shelley L. Martin (1995). *Therapeutic recreation in the nursing home.* State College, PA: Venture Publishing.

Callahan, Michael J. and J. Bradley Garner (1997). *Keys to the workplace: Skills and supports for people with disabilities.* Baltimore, MD: Paul H. Brookes Publishing.

Finlay, Linda (1993). *Groupwork in occupational therapy.* Cheltenham, UK: Stanley Thornes, LTD.

Haas, Karen L. and Robert McCartney (1996). The therapeutic quality of plants. *Journal of Therapeutic Horticulture* VIII: 61-67.

Hagedorn, Rosemary (2000). *Tools for practice in occupational therapy: A structured approach to core skills and processes.* London,UK: Churchill Livingstone.

Kemp, Jerrold E., Gary R. Morrison, and Steven M. Ross (1998). *Designing effective instruction.* Upper Saddle River, NJ: Prentice-Hall.

Lamport, Nancy C., Margaret S. Coffey, and Gayle I. Hersch (2001). *Activity analysis & application.* Thorofare, NJ: SLACK, Inc.

Murrey, Gregory J., Ann Wedel, and Jeff Dirks (2001). A horticultural therapy program for brain injury patients with neurobehavioral disorders. *Journal of Therapeutic Horticulture* XII: 4-8.

Schwebel, Andrew J. (1993). Psychological principles applied in horticultural therapy. *Journal of Therapeutic Horticulture* VII: 3-12.

Toseland, Ronald W. and Robert F. Rivas (2001). *An introduction to group work practice* (Fourth edition). Needham Heights, MA: Allyn and Bacon.

Chapter 5

Documentation: The Professional Process of Recording Outcomes

Sarah Sieradzki

INTRODUCTION

Documentation of horticultural therapy outcomes is an essential part of professional practice. This chapter will discuss the reasons documentation is important and offer general guidelines on ensuring its effectiveness and professionalism. Specific aspects of documentation will also be discussed, including:

- Types of documentation used when providing horticultural therapy to individual clients
- Detailed information on assessments, intervention or treatment plans, progress or observation notes, and discharge summaries
- Ways to make written and verbal communication more effective
- Types of documentation used when providing horticultural therapy to groups.

PURPOSE AND IMPORTANCE OF DOCUMENTATION

Whether a horticultural therapy program is in a therapeutic, vocational, or social setting, documentation provides a major difference between a professional program and a nonprofessional one. The following list offers purposes of documentation:

Horticultural Therapy Methods
doi:10.1300/5676_05

- Meets professional responsibility and obligations
- Provides a written record of services and treatment process
- Communicates with treatment team or other disciplines
- Monitors client progress and demonstrates outcomes
- Fulfills requirements by some regulatory or accrediting bodies
- Supplies information to obtain reimbursement for services
- Provides data for research, process monitoring, and quality assurance purposes
- Helps horticultural therapy gain credibility and recognition
- Provides information for program evaluation, adaptation, and improvement

Documentation is the written record of the entire process of assessing client function, identifying needs, establishing goals, monitoring and reporting client progress, adapting the program in response to feedback to improve client progress, and reporting positive outcomes and results of horticultural therapy. Written and verbal communication with the treatment team is essential to provide valuable information about the client's progress. Effective communication also serves to enhance the credibility and recognition of horticultural therapy as a valuable profession.

In horticultural therapy programs, documentation is an integral part of the treatment process. See Table 5.1. It creates a written medical/educational and legal record of treatment procedures, and provides important information about the client to other disciplines involved. Detailed information about documenting each step of the therapeutic process will be provided later in this chapter.

In many health care, rehabilitation, and educational settings, documentation is required by local, state, and federal laws. Knowing the specific rules of all the regulatory and accrediting bodies in each particular setting is vital to ensuring that documentation meets standards and requirements. These regulatory bodies may include the Joint Commission on Accreditation of Healthcare Organizations (JCAHO), Commission on Accreditation of Rehabilitation Facilities (CARF), Medicare, Medicaid, and state departments of special education, health, and mental health. Regulations are put in place to protect consumers by creating strict professional standards, to insure quality of care, and to justify the expense of reimbursing treatment. Ethical practice requires taking the time and effort to fulfill documen-

TABLE 5.1. Examples from Practice: Documentation in Various Types of HT Programs

Type of HT Program	Program Model	Aim of HT Program	Intent of Documentation
Therapeutic	*Medical* Settings may include: Hospital, skilled nursing facility, long-term care facility, rehabilitation center, pain management program, mental health setting, outpatient clinic, group home, special education setting, substance abuse/addictions recovery program, hospice	Management or recovery from an illness or injury Improved quality of life and functional independence	Demonstrate how horticultural therapy activities help client meet treatment goals Provide information to other disciplines or treatment team Meet requirements of regulatory bodies
Vocational	*Habilitation/Rehabilitation* Settings may include: Special education settings, MRDD programs, sheltered workshop, rehabilitation center, work hardening program, correctional facility (prison), programs for youths at risk, traumatic brain injury program	Increase specific vocational skills to maximize employability Maximize functional independence and ability to work with others	Demonstrate how horticultural therapy activities help client meet vocational training goals Provide information to other disciplines Meet requirements of regulatory bodies
Social and Wellness	*Community Group* Settings may include: Community gardens, arboreta, older adult day care, community wellness programs, garden clubs, mental health day treatment program, social service agency, support group, dementia programs	Enhance personal satisfaction, quality of life, sense of well-being Increase ability to improve wellness, prevent illness, or cope with health issues	Demonstrate how horticultural therapy activities can enhance the wellness of individuals/groups and improve quality of life

tation obligations. Failure to meet documentation regulations in a timely, comprehensive manner can have serious consequences legally and financially. Insurers may also require documentation before providing reimbursement for horticultural therapy services. Attention to detail in documentation will demonstrate the quality of care being given.

In addition to the types of documentation addressed in this chapter, documentation also establishes a data record that can be collected and used for other purposes. For example, many facilities utilize process-monitoring procedures for quality assurance and performance improvement. Documentation may also serve to provide valuable information for program evaluation aimed at improving the services and outcomes of a particular horticultural program or site. Furthermore, written documentation data can be collected, measured, described, and reported for research purposes to increase the body of evidence regarding the value and benefits of horticultural therapy. Outcome studies such as this improve the professional image and visibility of the profession.

GENERAL GUIDELINES FOR HORTICULTURAL THERAPY DOCUMENTATION

Because of differing target audiences, populations, regulations, and types of horticultural therapy programs, each setting is required to tailor its documentation to meet specific needs. To maximize its effectiveness, however, all documentation benefits from following certain guidelines as described in the following list.

- Timely
- Signed and dated
- Organized and easy to read
- Site or population specific
- Clear and understandable
- Legible and neat
- Professional and well written
- Meaningful and relevant
- Accurate and specific
- Descriptive and objective
- Kept confidential
- Directed at target audience

These suggestions are also helpful in the process of designing the particular format of documentation to be used in an individual horticultural therapy setting.

No matter what recording method is selected, documentation must be completed and shared in a timely manner for it to be meaningful to the target audience. In health-care and educational settings, there may be regulations that set time parameters for when different types of documentation must be completed. It is important to know the unique requirements of each horticultural therapy setting so that documentation can be completed by required deadlines. It is easy to become caught up in the more interesting and creative tasks of program development and providing horticultural therapy to a given population, and not leave adequate time for essential documentation. It is critical to build documentation time into the daily schedule so it can be completed in a timely manner.

Keep in mind that documentation creates a legal record of treatment provided, so it is important to sign and date each document. When signing a document, place professional credentials after the signature.

Keep documents well organized and easy to read. It is generally helpful to use as simple a format as possible. When writing, be direct, concise, and to the point. Include only important observations and pertinent details, and keep related points together. Add emphasis to the most significant points by stating them first, and then follow through in order of importance.

Use a format that is site or population specific and determine contents based on needs of target audiences. When developing the format for documentation, it is crucial to meet the needs of the target audience that will read the notes. This may include regulatory and accrediting bodies, insurers that are asked to reimburse services, other disciplines that form the treatment team, family members, and in certain instances, the client. The format can be designed as simply or as comprehensively as required for the needs of the program. It is also helpful to consider the time that will be allotted for documentation purposes. Checklists, charts, surveys, or graphs may be ways to simplify writing tasks rather than using more time-consuming narrative notes. Networking with other horticultural therapists in similar settings may provide ideas for easy and effective documentation formats.

Strive for clarity when documenting horticultural therapy procedures and client progress. Avoid unnecessary jargon or other easily misunderstood terms. Do not use confusing abbreviations unless a key is readily available on the document. Ambiguous terms such as "fair" or "poor" are usually not helpful as a measurable description.

Writing should be legible and neat. Keep in mind that the appearance of a document says a lot about the time and care taken to create it. High-quality documentation is a tangible representation of a quality horticultural therapy program. Black pen is generally considered best for professional documentation, as it is most readable. If a writing error occurs, the legal way to correct it is to cross it out with one line, and add the notation "Error" along with the writer's initials.

Documentation should be professional and well written. Always use correct spelling and punctuation when recording horticultural therapy documentation. When writing narrative notes, use complete sentences, correct sentence structure, and good grammar. Documentation is usually done in the third person—using "staff" or "therapist" instead of the words "I" or "me."

Write to express clear ideas rather than to impress someone with jargon or flowery language.

Make documentation meaningful and relevant. Document the most important information that will be useful to the target audience. This usually means focusing on the client's functional abilities and progress toward established treatment goals. Horticultural therapy has a unique perspective in observing clients actually performing an activity. Thus the therapist may have pertinent information to communicate regarding the task performance and organizational skills of the client.

Written or verbal communication should be accurate and specific. It is helpful to be as specific as possible when documenting. Vague reporting will cloud the significance of client and activity observations. Wherever possible, use measurable terms and record factual information. Bias has no place in professional documentation. It is crucial to be truthful even if the horticultural therapy procedures were not as satisfactory as desired, or when goals remain unachieved.

Be descriptive and objective in reporting. It is usually best to be factual in reporting observations of what actually occurred during horticultural therapy sessions. Be careful not to make assumptions or use inferences when reporting—staff commentary, which indicates bias or personal opinion, is inappropriate. Value judgments such as "nice," "pleasant," and "well-behaved" are also not useful or appropriate. It is helpful to use measurable terms (see Appendix A for examples) whenever possible, and use descriptive language to tell the story of what occurred during horticultural therapy sessions. Write as holistically and comprehensively as possible given time and space restraints. Do not use ambiguous terms that are open to misinterpretation.

All documentation about clients should be kept confidential. It should be shared only with necessary treatment team or other staff members. If documentation will be shared with insurers or other agencies, permission must be obtained from the client or the client's legal guardian. Many local, state, and federal regulators have strict confidentiality standards which must be followed; failure to do so may have legal consequences.

INDIVIDUALIZED DOCUMENTATION

Considerations

Therapeutic and vocational types of horticultural therapy programs generally require documentation to be completed regarding each individual client following every step in the therapeutic process. See Table 5.2. Some documentation may be mandated, such as Individual Education Plans (IEPs) in special education settings. Some

TABLE 5.2. Tips for Practice: Documentation and the Therapeutic Process

Steps in the Therapeutic Process	Documentation Which May Be Applicable
Initial assessment to determine client's abilities and identify needs	Initial evaluation report
Establishment of treatment goals and objectives as part of intervention plan	Written treatment plan that includes long-term goals, short-term objectives, how often specific interventions will be used, target dates for goal achievement, and who is responsible to provide treatment
Intervention through horticultural therapy activities (individual or group setting)	Group or individual treatment session observation notes, including record of adapted tools, techniques, and methods used
Monitoring and reporting progress through ongoing reassessment	Progress notes, including goal updates at prescribed intervals
Modification of treatment to maximize progress	Adaptation or addition of goals, and progress notes providing record of modifications and resulting progress
Reporting outcomes/results of treatment	Discharge summary, including goal outcomes

settings may have set formats for documentation; in other settings, horticultural therapy staff may be responsible for designing their own forms for documentation. Some settings may utilize computer-generated documentation, while others will rely on written and/or verbal reporting.

Types of Individualized Documentation

Each step in the therapeutic process may require a different type of document to capture the necessary information for reporting horticultural therapy process and outcomes. Each type of document is described in detail in the following chapter section, including the document's purpose, methods and procedures, timing, possible formats, and content areas. Examples of documents are found in Appendixes C and E.

Initial Assessments

Purpose. An initial assessment is conducted to determine the present level of functioning of the client and to identify treatment needs. The specific format, method, and content of the initial assessment are dependent on the site and population of the facility, as well as on requirements by regulatory and accrediting bodies.

Methods and procedures. There are a variety of ways to determine the present level of function of the client and needs for treatment, including self-reporting surveys, performing standardized evaluations, or using observations of the client performing evaluative tasks. A situational assessment is the observation of a client's behavior during a horticultural activity. A work sample is a standardized highly structured concrete task completed within a specific timeframe. Measurements are taken regarding the client's accuracy and amount of work completed during the timed task.

Timing. The initial assessment is completed as soon as possible after the client is referred and prior to the development of treatment goals and the intervention plan. Some settings have mandated time frames for completion of the evaluation process.

Possible formats. Initial assessments can take various formats from simple checklist self-surveys to more comprehensive charts. In some settings where ongoing reassessments are done to determine the client's progress, the initial assessment form might include additional columns in which to document these periodic updates. See Appendix C for examples of assessment formats.

Content areas. Most initial assessments identify a client's strengths and abilities as well as limitations and needs. Specific content varies depending on the requirements of the site, population, regulatory bodies, and target audiences. In vocational settings, most initial assessments include information about a client's cognitive function, task performance skills, cooperative attitude and behavior, and prevocational skills. In health-care settings, assessment content may include physical neuromuscular function such as fine and gross motor skills, range of motion, muscle strength, endurance, and sensation. It may also include detailed information about cognitive function, task performance skills, communication skills, leisure skills and interpersonal skills. See Appendixes C and D for examples of assessment content.

Intervention Plan or Treatment Plan

Purpose. A treatment or intervention plan outlines the specific way horticultural therapy staff (and other involved disciplines) meet the identified needs of the client. When done well, it provides a road map for the interventions that will follow, and sets parameters for goal achievement and eventual discharge from treatment.

Methods and procedures. In some health-care and educational settings, a multidisciplinary treatment plan is mandated which includes timing, the basic format, and content. The entire team may work together to develop the intervention plan. For example, in special education settings, state or federal laws may set requirements for an IEP as shown in Exhibit 5.1. In an acute care hospital setting, the JCAHO may require specific timing, contents, and procedures for a client's Interdisciplinary Treatment Plan (ITP). In a rehabilitation setting, CARF may set the standards for the plan of care. In some settings, the horticultural therapist may develop the treatment plan individually. Wherever possible, the client is included in the identification and establishment of goals. This will lead to client-centered practice in which the top priority goals are most meaningful to the client, which results in increased motivation during treatment. See Table 2.1 in Chapter 2 for information about who is typically included on treatment teams.

EXHIBIT 5.1.
Tips for Practice: Components
of an Individual Education Plan

The federal government has mandated some specific components for Individualized Education Plans for school-age students with an identified disability. State and local boards of education may require additional content for IEPs as well. In general, the customary IEP components are:

- A description of the child's present levels of educational and/or functional performance, including academic achievement, pre-vocational and vocational skills, psychomotor skills, and self-help skills
- An indication of how the child's disability impacts his or her educational performance
- Annual goals which describe what the child can reasonably be expected to achieve during the current school year, and specific criteria for measuring goal accomplishment
- Short-term objectives which provide measurable steps toward achievement of the annual goals, and how the parents will be regularly informed of the child's progress
- Identification of (and justification for) the particular special educational and related services needed by the child along with any specific accommodations (such as aids, modifications, and supports) required in the classroom or during standardized testing to meet that child's unique needs
- The extent to which the child will participate in regular education programs with nondisabled children to help ensure the least restrictive placement for the child
- Specific information regarding services to be provided including beginning and ending dates, frequency, location, and duration of services, and who will be responsible for providing them
- For children aged fourteen and older, identification of transitional services (such as linkage to other agencies and interagency responsibilities) needed to help the child reach post-school goals such as post-secondary education, vocational training, independent living, community participation, and employment (including supported employment).

Timing. The intervention or treatment plan is written as soon as possible following the initial assessment, and prior to the start of actual therapy sessions.

Possible formats. The treatment plan format may be mandated by a regulatory or accrediting agency (see "methods and procedures" above), or developed at the site of horticultural therapy practice. In some settings, the treatment plan is an integral part of the initial assessment format, while in other sites it is an independent document.

Content areas. The intervention plan generally includes the client's identified needs or problem areas, long-term goals, short-term behavioral objectives, specific interventions to be used (group or individual treatment sessions), how often and how long treatment will occur, which specific disciplines or persons are responsible to carry out treatment procedures, and target dates for goal achievement. Creating effective treatment goals and objectives is key to easy measurement of outcomes later in the process. A simple technique for writing effective goals is to use the SMART method outlined here. SMART goals are:

> **S**pecific: States exactly what the client will accomplish
> **M**easurable: Has the number of items or times something will be done
> **A**ction oriented: Tells how the goal will be achieved—stated using a verb
> **R**ealistic: Enough to be challenging but not too difficult
> **T**ime based: Has an end point or deadline

An effective goal is specific, measurable, action oriented, realistic, and time based (has a deadline). A SMART goal succinctly states what the client will accomplish in a particular way by a certain deadline. SMART goals can be written for both individuals and groups who receive horticultural therapy services. At the conclusion of the therapy session(s), the therapist and client can quickly and easily see if the goal has been accomplished, and documentation of progress will be simpler.

Knowing what to measure is often the most difficult part of goal writing and documentation. See Appendix A for a broad array of horticultural therapy short-term goals, examples of treatment activities, and what to measure to show treatment effectiveness in various types of horticultural therapy programs. Long-term goals are often to improve functional abilities, independence, or quality of life of the client.

Progress Notes

Purpose. Progress notes are the ongoing record of horticultural therapy treatment sessions and changes in the client's functional status in response to this treatment.

Methods and procedures. In some horticultural therapy programs, observation notes may be written following each individual or group therapy session. In other settings, progress notes are performed periodically, often on a weekly basis.

Progress can be measured in either quantitative or qualitative terms. When using a quantitative method, progress is expressed in a countable, comparable, quantifiable way—usually using a checklist, graph, or chart of time amounts, percentage of accuracy, or number of completed tasks. When using a qualitative method, the progress note demonstrates changes in client behavior or function through a descriptive characterization such as a narrative or case study.

There are several ways to document progress: a change in the level of assistance the client needs to perform a specific task, a change in response to treatment (example—increased consistency of attendance), development of a new skill or ability, or learning a new compensatory technique.

It is also important to document temporary setbacks, such as slow (or complete lack of) progress due to pain, medication change, illness or medical complications, or external circumstances such as family or scheduling issues. When setbacks are identified, it is important to state the perceived reason and the plan of action to enhance future progress.

Timing. In some settings, the frequency of progress notes is mandated, while in others frequency will be determined by time and space constraints. A progress note is performed weekly in many settings. For long-term settings, such as rehabilitation centers, sheltered employment, or residential programs, the interval between progress notes may be longer.

Possible formats. Documentation of progress may take the form of checklists, rating scales, or narrative notes. If the facility uses a problem-oriented medical record, the format used by all disciplines may be the SOAP format described here. SOAP documentation includes the following:

Subjective information from the client
Objective observations of staff and clinical findings
Assessment of progress toward established treatment goals or interpretation of reassessment data
Plan for further treatment including any potential changes to enhance progress

In other settings, graphs, surveys, or charts may be used. For example, in vocational settings, documentation may include detailed information about performance on the steps of a horticulture task or demonstration of pro-social behaviors. When selecting or developing a format, emphasis should be placed on including the most comprehensive amount of information in the simplest and most concise fashion to expedite documentation duties. See Appendixes C and E for examples of forms used for recording client progress.

Content areas. Progress notes will often include a record of attendance, level or quality of participation, description of treatment sessions, changes in client behavior or function, other relevant observations, the therapist's interpretation of the reassessment data, an update of progress toward goal achievement, and the plan for further horticultural therapy involvement. The future plan may include the modification of goals or adaptation of horticultural therapy procedures to enhance client progress, or the identification of new goals if the first objectives have been accomplished.

Discharge Summary

Purpose. The discharge summary is a report that recaps the entire treatment process and records the final reassessment and update of goal achievement. It is generally a brief synopsis of outcomes; however, it needs to be as comprehensive as possible within a concise and simple format. Discharge summaries are often the document most read by insurers to determine if services will be reimbursed and by regulators to ensure quality of care.

Methods and procedures. Some facilities use a separate format for a discharge summary while others write a more detailed final progress note. Most discharge summaries document the amount of goal achievement shown as well as recommendations for follow-up care.

Timing. The discharge note should be written as soon as possible following the discontinuation of horticultural therapy services. In some settings, regulators mandate the time frame.

Possible formats. The discharge summary document can be a narrative note, a checklist of progress, a self-evaluation such as a satisfaction survey, or a graphic representation of results such as a chart or graph of progress.

Content areas. The discharge summary generally contains information about attendance, the number of sessions the client experienced, a reassessment of the client's functional abilities and update of goal achievement, a review of any adaptive tools/methods/techniques used, any needed interpretation of results, and recommendations for further follow-up care. See Appendix C for an example of discharge summary contents and format.

Verbal Reports

Purpose. Horticultural therapists are often called upon to give verbal reports of client progress in team meetings and conferences. Similar principles of good communication apply to verbal reporting as to written notes.

Methods and procedures. If possible, prepare ahead of the meeting for what needs to be included in the verbal report. Just as in written records, emphasize the most significant information by reporting it first, and then follow in descending order of importance. Be succinct, and only report pertinent and meaningful observations. Many team conferences are held as a forum, combining information from numerous sources, to determine the future course of treatment for a client or for group problem-solving purposes.

Timing. Conferences may be held at any time throughout the treatment process; however, in certain settings team meetings may have a mandated frequency.

Possible formats. Verbal reports may be given as part of an IEP meeting in an educational or vocational setting, or as part of an ITP meeting in a medical or rehabilitation setting. Meetings may also be held with administrative staff, the client, family members, or a case manager from another involved social service agency. In community horticulture programs that do not necessarily include a full treatment planning process, verbal reports may be the primary means used to communicate the progress of and development of individual participants.

Content areas. Often the individual responsible for scheduling the conference provides an idea of what information or problem will be discussed. This allows the horticultural therapist to prepare specific examples of goal progress or pertinent adaptive methods to share with others. Having the opportunity to prepare ahead of time may even allow for the development of graphic or written supplementary materials to provide more comprehensive information to the group. Information content may vary, but usually will include details regarding client response, behaviors, problem areas, and progress toward goals.

GROUP DOCUMENTATION

Because horticultural therapy is often provided to groups of clients, some programs may utilize group documentation in addition to, or in place of, more highly detailed individualized documentation. Group documentation will usually be employed when the program uses group goals rather than individualized objectives for each client. Group documentation may also be used when the horticultural therapy group meets infrequently—monthly, for example. Although documentation about individual clients within a group may also be recorded, the methodology needs to be simpler and less time consuming than that described in the previous section when recording about a larger number of clients.

In this section, group documentation will be described including its purpose, methods and procedures, timing, possible formats, and content areas.

Purpose. Social or wellness types of horticultural therapy programs often involve providing activities to meet group goals. Group documentation will record the needs, goals, and outcomes of group horticultural therapy programming.

Methods and procedures. Documentation of group horticultural therapy programs follows a similar pattern of steps in the therapeutic process as just described in individualized documentation. An assessment procedure takes place prior to the initiation of horticultural therapy sessions to determine the needs and goals of group members. A report of observations of each group session may be desirable if time allows. Reviewing this information assists the horticultural therapist in assessing ongoing progress and client satisfaction. The therapist can then make any needed activity modifications to enhance the program for group members. At the conclusion of the predetermined number of group sessions, a final summary report outlines goal outcomes and any other pertinent information about the horticultural therapy experiences of the group.

Timing. When programs are offered with a beginning and end, assessment and goal development should take place in the first group session and reassessment should occur in the last horticultural therapy session of the group program. For ongoing groups, periodic documentation is performed.

Possible formats. Simpler versions of individual client documentation may be utilized to collect and record group session information. Using a client survey as a preassessment, and then again as a postassessment, will provide a comparative method of self-evaluation. The use of rating scales with results that can be compared visually using a chart or graph is another easy method of documentation. See Appendix C for an example of a group documentation format.

Content areas. Contents of group documentation may be similar to that of individualized documentation or it may be entirely different depending on the needs and goals of the group program. Social and wellness types of horticultural therapy programs may emphasize personal satisfaction and quality of life issues rather than specific skill development or functional capability. See Appendix A for sample SMART goals, treatment activities, and ideas for measuring progress for social and wellness types of horticultural therapy programs.

SUMMARY

This chapter described the customary methods and content of individual and group client documentation used in professional horticultural therapy practice. Performing documentation effectively is crucial for the credibility and future growth of the profession.

SUGGESTED READINGS

Austin, David R. (1991). *Therapeutic recreation: Processes and techniques.* Champaign, IL: Sagamore Publishing.

Borcherding, Sherry (2000). *Documentation manual for writing soap notes in occupational therapy.* Thorofare, NJ: Slack, Inc.

Davis, William B., Katie E. Gfeller, and Michael H. Thaut (1992). *An introduction to music therapy: Theory and practice.* Dubuque, IA: Wm. C. Brown Publishers.

Ozer, Mark, Otto D. Payton, and Craig E. Nelson (2000). *Treatment planning for rehabilitation: A patient-centered approach.* New York: McGraw-Hill.

Simson, Sharon P. and Martha C. Straus (Eds.) (1998). *Horticulture as therapy: Principles and practice.* New York: The Haworth Press, Inc.

Appendix A:
Goals, Activities, and Measurements

SMART GOALS, ACTIVITIES, AND MEASUREMENT
FOR THE THREE TYPES OF HT PROGRAMS

Following are examples of SMART goals that might be applied in various HT settings. For each type of program—vocational, wellness, and therapeutic—potential settings and diagnoses are specified, with ideas for goals, activities and what to measure.

Horticultural Therapy Methods
© 2006 by The Haworth Press, Inc. All rights reserved.
doi:10.1300/5676_06

Vocational HT Programs

Vocational Training or Rehabilitation:

- *Settings:* Work hardening clinics, sheltered workshop for MR/DD clients, TBI program
- *Typical Diagnoses and Other Types of Participants:* Developmental disabilities, mental retardation, youth at risk, traumatic brain injury, correctional facility (prison) population, hand injuries, back injuries, special education, spinal cord injuries

Short-Term Goal/ Objective	HT Activity	What to Measure
Client will remember and follow 3-step directions with 80% accuracy without visual or verbal cues after 2 treatment sessions	Planting stem tip cuttings to be grown for sale	Comparison of trials of same 3 step directions over time to determine accuracy percentage
Client will add prices of sale plants and give correct change with 80% accuracy.	Performing cashier job at plant sale	Determine accuracy percentage of several trials
Client will demonstrate accurate counting skills from 1-12, 100% of the time after 5 learning trials.	Each trial will consist of placing 12 pots in flat to be used for potting up plant divisions	Determine accuracy of 12 pots per flat in several trials
Client will clean worksite after task completion with no verbal reminder after 3 introductory learning trials.	Use of whisk broom and dustpan to clean up potting mix from table and floor, and then washing hands with soap and water	Chart number of times per week that client proceeds with appropriate cleanup tasks at conclusion of work session without verbal reminder

Social and Wellness HT Programs

Community Wellness Programs:

- *Settings:* Botanical garden or arboretum, community health center, support group, nonprofit health agency (such as National Multiple Sclerosis Society, American Heart Association, American Cancer Society, Alzheimer's Association, National Alliance for the Mentally Ill), group homes, day treatment programs
- *Typical Diagnoses and Other Types of Participants:* Individuals with MS, heart disease, cancer, arthritis, or other chronic or debilitating illness; caregivers of individuals with chronic illness or disability; individuals who are dealing with grief and loss.

Short-Term Goal/ Objective	HT Activity	What to Measure
During first support group session, client will set 2 SMART goals to positively impact physical or emotional health.	Potting cuttings of plant that best represents how client would like health to be in the future (selected from variety of plants). HT activity leads to small group discussion of SMART goals.	Client's written SMART goals and action plan of ways to positively impact health.
Client will identify 2 positive ways to better cope with illness and stressors by end of 3 support group sessions.	Creating potpourri using dried flowers, dried fragrant herbs, and spices. Activity leads to discussion of use of fragrance for relaxation and other methods of self-care.	Client's written or verbal identification of 2 potentially helpful coping strategies.
Client will discuss a way to better cope with grief and loss by conclusion of 2 support group sessions.	Arranging pressed flowers to be put in a frame, using Victorian meanings of flowers and leaves to describe loved one. Discussion of grief and loss will accompany activity.	Client's written or verbal identification of a way to safely express feelings and better manage emotions as part of grief process

Therapeutic HT Programs

Physical Rehabilitation:

- *Settings:* Rehabilitation center, skilled nursing facility, home health care, orthopedic/sports medicine program, work hardening program
- *Typical Diagnoses:* CVA, hip fractures, orthopedic/sports injuries, burns, TBI, spinal cord injuries, MS, neurological disorders, hand injuries, repetitive use injuries such as carpal tunnel or tennis elbow, joint replacement surgeries

Short-Term Goal/ Objective	HT Activity	What to Measure
Client will extend horizontal reach by 10 inches forward and 6 inches to each side after 3 treatment sessions.	Reaching forward and side to side to plant a tray of seeds	Length and arc of forward reach in inches
Client will tolerate standing for 30 minutes after 2 sessions.	Potting up plant cuttings while in standing table	Length of standing tolerance time from session to session
Client will improve muscle strength from poor to fair and length of endurance for upper extremity tasks up to 20 minutes without a break after five treatment sessions.	Filling pots with soilless mix as part of group assembly line potting task	Muscle testing or increase in weight of pots with soil. Length of time client can work on task before tiring and needing a break
Client will improve pincer grasp by increasing number of seeds sown by 100% after 3 treatment sessions.	Planting seeds of various size by using pincer grasp	Speed and accuracy of picking up and planting seeds (comparison of timed trials)
Client will show improved eye-hand coordination by improving speed and accuracy of watering task by 50% over 3 treatment sessions.	Watering trays of small pots using squirt bottle or watering bulb	Speed and accuracy of watering task (comparison of several timed trials)

Pediatric Health Care:

- *Settings:* Acute pediatric hospital unit, special child day treatment program, school-based special education setting, pediatric rehabilitation center
- *Typical Diagnoses:* Cerebral palsy, autism, MR/DD, burns, fractures, muscular dystrophy, accident trauma, poisonings, brain injuries, surgical procedures, childhood cancers

Short-Term Goal/ Objective	*HT Activity*	*What to Measure*
Client (receiving chemotherapy) will explore at least 2 ways to cope with impending hair loss to reduce impact of feeling different.	Creating "grass goon" creature with facial features to try out various grass "hairstyles" leading to small group discussion of issue	Client's identification of 2 ways to cope with embarrassment of impending hair loss following chemotherapy treatment
Client will demonstrate improved eye-hand coordination and pincer grasp by increasing accuracy of seed sorting by 80% in 3 treatment sessions.	Sorting seeds (of a variety of size and shape) into small containers, then choosing seeds to plant in own peat pots to take home to be planted	Measuring timed trials for accuracy of seed sorting
Client will identify 2 ways to contribute to own recovery and maintain improved health/wellness after discharge in 3 treatment sessions.	Adopting a plant that needs frequent care to experience nurturing another living thing (rather than always being recipient of care), then participating in discussion of best ways to care for plant and self in the future	Client's identification of 2 ways to better care for own health/wellness after discharge from hospital setting
Client will cooperatively practice gait training in 2 physical therapy sessions (working cooperatively with horticultural therapist to co-treat client).	Planting activity with materials in various places around the room	Client walking around room to obtain various supplies for activity with few or no complaints

Mental Health:

- *Settings:* Inpatient acute care behavioral health unit, community mental health center, partial hospitalization program, group home, home health, addiction recovery
- *Typical Diagnoses:* Major depressive disorder, bipolar disorder, anxiety and panic disorders, schizophrenia, personality disorders

Short-Term Goal/ Objective	HT Activity	What to Measure
Client will explore 1-2 healthy and productive leisure pursuits for use post-discharge during 2-3 treatment sessions.	Introduction to houseplant varieties and care, or development of small container garden to take home	Client's identification of 1-2 leisure pursuits to be used at home post-discharge with detailed information on what, when, how, and where
Client will demonstrate cooperation and positive social skills in 3 treatment groups prior to discharge.	Horticultural tasks such as seed sowing and planting stem tip cuttings to take home	Client's ability to cooperate in group with peers without disruption or agitation in 3 group sessions
Client will demonstrate improved communication skills by verbally requesting materials and seeking help to clarify directions in 2 treatment sessions prior to discharge.	Planting edible flower salad container garden or potpourri-making activity (tasks that involve use of variety of materials and have multiple-step directions)	Client's ability to ask for materials and verbally request assistance or clarification of instructions in 2 treatment sessions
Client will identify 2-3 positive coping strategies to better manage illness and stressors post-discharge.	Plant division activity in small group leading to discussion of meeting needs of plant and self to promote positive growth	Client's identification of 2-3 methods of coping to better care for meeting own needs post-discharge
Client will demonstrate adequate behavioral self-control to complete one 30-minute horticultural task without agitation, aggressive behavior, or verbal outburst.	Horticultural tasks such as grooming dead leaves off scented geranium plants, and potting up offshoots from succulent plants	Client being able to tolerate 30-minute tasks including staff instructions/redirection without angry verbal outburst or agitated behavior

Pain Management:

- *Settings:* Pain management clinic, home health care, rehabilitation center, skilled nursing center, orthopedic/sports medicine program, hospice
- *Typical Diagnoses:* Arthritis, back injuries, shingles, cancer, post-surgical pain, fractures, repetitive use injuries, migraines, degenerative spinal discs, burns

Short-Term Goal/ Objective	HT Activity	What To Measure
Client will demonstrate reduced intensity of pain perception during HT activity by two points on pain scale after 2 treatment sessions.	Potting up scented geranium cuttings, or cutting roses and separating petals to dry for potpourri-making purposes	Change in perceived pain intensity using 1-10 pain scale (comparison of trials)
Client will identify 2 new coping strategies to better manage pain after 2 treatment sessions.	Planting an edible flower salad garden in a large container in small group	Client's identification of 2 new pain management strategies

Hospice:

- *Settings:* Residential hospice unit, home health care
- *Typical Diagnoses:* Terminal cancer, congestive heart failure, degenerative neurological disorders

Short-Term Goal/ Objective	HT Activity	What to Measure
Client will experience reduced perception of pain by two points on pain scale during and immediately following an HT treatment session.	Arranging bouquet of cut flowers to be placed in client's room or given to loved one	Change in perceived pain intensity of 2 points on 1-10 pain scale during HT activity and for 30 minutes afterwards
Client will express sense of peace and closure regarding impending death in one HT treatment session.	Discussion of type of tree to be planted to leave legacy for loved ones and discussion of end-of-life issues	Client's verbal expression of readiness for natural death or sense of peace in having closure with loved ones

Specialized Dementia:

- *Setting:* Alzheimer's unit in skilled nursing facility, geriatric adult day care program, home health care, inpatient geropsychiatry unit
- *Typical Diagnoses:* Alzheimer's dementia, multi-infarct dementia, Pick's disease, Huntington's disease, Crutchfield-Jacob's disease, traumatic brain injury, CVA

Short-Term Goal/ Objective	HT Activity	What to Measure
Client will demonstrate orientation to self during HT treatment activity after 3 treatment sessions	Care for houseplants brought to unit on cart along with reminiscence discussion of plants, gardening	Client will verbally express a memory of own past gardening experiences or identify two types of plants
Client will demonstrate reduced agitation and ability to socialize appropriately with peers for 15 minutes during HT treatment activity.	Division and planting of fragrant herbs into container gardens for unit courtyard or planting heirloom plants in containers	Client will display no disruptive behavior and will verbally address staff or peer in appropriate manner for 15 minutes during activity
Client will follow 2-step directions with verbal cues for 15 minutes in 3 HT treatment sessions.	Arranging bouquet of fresh or dried flowers to place in client's room (nontoxic plants only)	Client's ability to follow 2-step directions for 15 minutes

(Contributed by S. Sieradzki)

Appendix B:
Activity Planning

Horticultural Therapy Methods
© 2006 by The Haworth Press, Inc. All rights reserved.
doi:10.1300/5676_07

HORTICULTURAL THERAPY ACTIVITY PLAN

Group Name: _____ Session Date:_____

Program Model Type: _____

Goals: _____
_____
_____
_____

Methods/Activity Summary: _____
_____
_____
_____

Supplies: _____
_____
_____
_____

Procedure: _____
_____
_____
_____
_____
_____
_____

Evaluation Method: _____
_____
_____
_____
_____

Review and Follow-Up Notes: _____
_____
_____
_____

(Contributed by K. Kennedy for the Horticultural Therapy Institute)

ACTIVITY: TERRARIUMS

Time Allowed: 1 Hour

Goals:

1. Maintain fine motor skills; develop/maintain cognitive skills.
2. Develop/maintain appropriate socialization skills

Objectives:

1. Participants will utilize fine motor skills for at least 10 minutes.
2. Participants will maintain alertness during 5-10 minutes of discussion time during session.
3. Participants will work cooperatively with others for duration of project.

Materials Needed:

- clear plastic "suitable for terrarium" containers
- pencils/pens for poking holes
- potting mix and trays or bags
- gloves
- aquarium gravel
- charcoal
- suitable plants (ivy, *Fittonia, Peperomia*)
- water, warm, cool
- scissors

Procedure:

1. Distribute terrarium containers to groups.
2. Cover bottom of terrarium with at least one inch of aquarium gravel.
3. Cover gravel layer with a thin layer of charcoal.
4. Mix potting mix with warm water in a resealable bag until evenly moist.
5. Cover gravel and charcoal with a 2" layer of potting mix pressed in firmly.
6. Determine where plants are to go (2 or 3 plants) and make hole for each.
7. Place cuttings in holes and firm potting mix around stems.
8. Water gently around each plant.
9. To remove soil from walls of container either pour a tiny amount of water down walls or take a tissue and wipe out.
10. Put on lid and place in an area with indirect light.

Planning Notes:

Points of Interest:

- Terrariums were first used by early discoverers as a way of keeping plants alive when bringing them back to their homelands from far across the oceans. Seawater couldn't be used and drinking water needed to be saved for the crew.
- Moisture from the soil & plants evaporates and rises to the surface where it condenses and returns to the soil like rain and the cycle continues.
- As long as the seal is tight, there is no need to add water in the future.
- Charcoal keeps the terrarium water "sweet." Did you know that charcoal is often used in water filters to create better tasting water?
- What makes a good terrarium container? Anything clear that can be sealed. Aquarium tanks, canning jars, soda bottles, giant pickle jars, giant water bottles, etc.
- Terrariums need to be kept out of direct sunshine, as plants can be burned from the intensity of the sun through the clear glass or plastic.

Adaptations:

- To encourage more socialization and cooperation among peers you can develop teams and have each team cooperatively create a terrarium.

(Contributed by P. Catlin)

RESOURCES FOR HT ACTIVITY IDEAS

Associations/Organizations

- American Horticultural Therapy Association (journals, information sheets) www.ahta.org
- National Gardening Association www.nationalgardening.com

Books

- Beverly, Deena. *Flowercrafts*. New York: Lorenz Books. 1997.
- Browning, Marie. *Making Glorious Gifts From Your Garden*. New York: Sterling Publishing Company. 1999.
- Bruchac, Joseph and Caduto, Michael J. *Native American Gardening Stories, Projects and Recipes for Families*. Golden, CO: Fulcrum Publishing. 1996.
- Cohen, Joy and E. Pranis. *Grow Lab Activities for Growing Minds*. Burlington, VT: The National Gardening Association; 1990.
- Gabaldo, Maria M., et al. *Health Through Horticulture: A Guide for Using the Outdoor Garden for Therapeutic Outcomes*. Glencoe, IL: Chicago Botanic Garden. 2003.
- Hewson, Mitchell L. *Horticulture as Therapy: A Practical Guide to Using Horticulture As a Therapeutic Tool*. Guelph, ON, Canada: Greenmor Printing Company LTD. 1994.
- Lovejoy, Sharon. *Sunflower Houses: Garden Discoveries for All Ages*. Loveland, CO: Interweave Press. 1996.
- Molen, Stephanie, et al. *Growth Through Nature: A Preschool Program for Children with Disabilities*. Sagaponack, NY: Sagapress, Inc. 1999.
- Moore, Bibby. *Growing with Gardening: A Twelve Month Guide for Therapy, Recreation, and Education*. Chapel Hill, NC: University of North Carolina Press. 1989.
- Newdick, Jane. *Herbal Gifts*. New York: Quadrillion Publishing, Inc. 1999.
- Ocone, Lynn and E. Pranis. *The National Gardening Association Guide to Kids' Gardening: A Complete Guide for Teachers, Parents and Youth Leaders*. New York: John Wiley & Sons, Inc. 1990.

Magazine Examples

- *Martha Stewart Living*
- *Parenting*
- *Sunset*
- *Women's Day*

Horticulture and Environment

Organizations/Associations

- American Horticultural Society www.ahs.org

- American Nurseryman Association
  www.amerinursery.com
- The National Arbor Day Foundation
  www.arborday.org
- Cooperative Extension Service: Located in most counties across the United States. Affiliated with land grant universities.
- EarthDay Foundation
  www.earthdayfoundation.org
- USDA Forest Service
  www.fs.fed.us

Books and/or Videos

- Reilly, Ann. *Park's Success With Seeds*. Greenwood, SC: George Park Seed Co., Inc. Rodale Publishing, Emmaus, PA. 1998.
- Squire, David. *Cassell's Directory of Scented Plants*. London: Cassell & Company. 2000.
- Sunset Books, www.sunset.com

Internet Horticulture Information Site Examples

- http://www.gardenweb.com
- http://www.plantcare.com
- www.garden.com

Television

- Home and Garden TV (HGTV)

Magazines

- *American Gardener* (American Horticultural Society)
- *Better Homes and Gardens*
- *Country Gardens*
- *Garden Gate*
- *Herb Companion*
- *Horticulture*
- *Organic Gardening*
- *Practical Gardener*
- *Sunset*

(Contributed by P. Catlin)

HOLIDAYS AND SIGNIFICANT CELEBRATIONS

January (flower: Carnation)

- New Year's Day
- Martin Luther King Jr. Day
- Chinese New Year

February (flower: Violet or Sweetheart Rose)

- Groundhog Day
- Lincoln's Birthday
- Valentine's Day
- President's Day
- Washington's Birthday
- Mardi Gras
- Ash Wednesday

March (flower: Jonquil or Daffodil)

- St. Patrick's Day
- Spring Solstice
- National Horticultural Therapy Week

April (flower: Daisy or Sweet Pea)

- April Fool's Day
- Daylight Savings Time begins (except AZ)
- Palm Sunday (sometimes in March)
- Passover
- Easter (sometimes in March)
- Earth Day (Apr. 22)
- Arbor Day (last Friday of April)

May (flower: Lily of the Valley)

- May Day (May 1st)
- Mother's Day
- Armed Forces Day
- Memorial Day

June (flower: Rose)

- Flag Day
- Father's Day
- Summer Solstice

July (flower: Cornflower)

- Independence Day

August (flower: Gladiolus)

September (flower: Aster)

- Labor Day
- Patriot Day
- Grandparents Day
- Rosh Hashanah (variable dates)
- Fall Solstice
- Yom Kippur (variable dates)

October (flower: Calendula)

- National Children's Day
- Columbus Day
- United Nations Day
- Halloween
- End of Daylight Savings Time

November (flower: Chrysanthemum)

- Election Day
- Veterans Day
- Thanksgiving

December (flower: Poinsettia or Narcissus)

- Pearl Harbor Day
- Hanukah
- Winter Solstice
- Christmas
- Kwanzaa

(Contributed by P. Catlin)

Appendix C:
Documentation

Horticultural Therapy Methods
© 2006 by The Haworth Press, Inc. All rights reserved.
doi:10.1300/5676_08

BASICS DOCUMENTATION

The following daily evaluation form was developed for the Horticultural Training Center at Father Flannigan's Home for Boys, "Boys Town" (now Girls and Boys Town), in Nebraska, The "BASICS" system of daily evaluation of youth employment focuses on positive advancement. Within a two-week pay period, each employee had the opportunity to earn their BASICS each day they worked. BASICS is an acronym for **B**eing on time, **A**ttire, **S**taying on task, **I**nstruction-following, **C**ondition of work, **S**ocial skills. Dependent upon percentage of excellence scores and competency points earned, employees were rewarded with raises.

Employee: J.S.　　　　　　　　　　Pay Period for the Weeks of:

M	T	W	Th	F	M	T	W	Th	F	Total Hours
4	4.5	4	3.5	2	4	4	4.5	4	4.5	39
+	+	+	+	+		+	+	+	+	B
.+	+	+	+	+	+	+	+	+	+	A
+	+	+	+		+	+	+	+	+	S
	+	+	+	+			+	+	+	I
+	+	+	+	+	+	+	+	+	+	C
+	+	+	+	+	+	+	+	+	+	S

Being on time	90
Attire	100
Staying on task	90
Instruction-following	70
Condition of work	100
Social skills	100
% of excellence (550/600)	91%

CONGRATULATIONS J.S.!!! You have successfully kept your percent of excellence above 90 percent for 6 pay periods. Your next check will reflect a \$.25 raise.

(Contributed by T. Kent Titze)

HANNA UNIVERSITY GEROPSYCHIATRIC CENTER
SITUATIONAL ASSESSMENT OBSERVATIONS

Client:

Session Activity: Date:

Observation:	Rating:	Comments:
Attitude Toward Task:		
Willing to engage		
Shows initiative and motivation		
Responsible with tools/plants		
Works till completion		
Follows safety precautions		
Punctual/effective time mgt.		
Understands task purpose		
Seeks help appropriately		
Emotional Control:		
Patient/delays gratification		
Tolerates frustration approp.		
Modulates own mood		
Shows impulse control		
Focuses on positives		
Manages anxiety		
Cognition/Task Performance Skills:		
Understands/recalls instructions		
Attends to task/topic		Attention span: _____ minutes
Follows multistep directions		
Follows appropriate sequence		
Aware of own errors		
Can organize own task		
Able to problem solve		

Functions best with: (circle)
Written instructions   Demonstration   Visual cues   Constant cuing
Physical assist   Hand over hand   Adaptive tools/techniques: (describe)

Communication/Interpersonal Skills:		
Socializes/tolerates peers		
Cooperates with others		
Accepts supervision/assist		
Shares tools/equipment/space		
Initiates interaction		
Responds in interaction		
Shows flexibility/tolerates change		
Behaves appropriately		
Shares own experiences/feelings		
Shows self-confidence		
Physical Skills/Abilities:		
Has understandable speech		
Has adequate hearing		
Has adequate vision		
Adequate sitting/standing balance		
Adequate gross motor skills		
Adequate fine motor/eye hand		
Adequate muscle strength		
Adequate endurance/energy level		

Independent mobility with: (circle)
Amulates independently   Crutches   Cane   Walker   Rollator
Wheelchair   Motorized wheelchair

Client strengths:
Client limitations:
Future needs:

Rating Scale:

1 = Able/independent
2 = Able with structure/supervision
3 = Able with minimal physical/cognitive assist
4 = Able with significant physical/cognitive assist
5 = Unable

(Contributed by S. Sieradzki for Hanna University Geropsychiatric Center)

GROUP SINGLE-SESSION OBSERVATION CHART

Deaconess Hospital of Cleveland Horticultural Therapy Program

Date: Group:

Session Activity:

CLIENT NAME:										
Followed and recalled multistep instructions										
Able to organize task in proper sequence										
Able to see and correct own errors										
Followed safety precautions										
Demonstrated emotional self-control										
Positive in interactions with staff and peers										
Shared own past feelings/ experiences										
Worked cooperatively with others										
Willing to engage for entire group session										
Showed flexibility/tolerated change										
Tolerated frustration appropri-ately										
Had adequate motor skills/ coordination										
Had adequate muscle strength/endurance										
Worked independently with minimal cues										

Additional Notes:

Therapist Signature: Date:

SAMPLE DISCHARGE SUMMARY

Deaconess Hospital of Cleveland Horticultural Therapy Program

Client Name: Medical Record Number:

Diagnosis: Date of Referral:

Discharge Summary:

Discharge Date:

Client participated in _____of_____ scheduled Horticultural Therapy sessions during this hospitalization. Level and quality of participation was:

Horticultural Therapy Goal Status:

	Achieved	Minimally achieved	Partially achieved	Not achieved	Comments:
Goal 1:					
Goal 2:					
Goal 3:					
Goal 4:					

Recommendations/Comments:

Therapist signature: Date:

(Contributed by S. Sieradzki for Deaconess Hospital of Cleveland)

CRAIG HOSPITAL HORTICULTURAL THERAPY REFERRAL
AND PATIENT PARTICIPATION

Patient:

Age: **DOI:**   **Room#**

TBI: **SCI:**
LOCF: Injury Level:
Comments:

Gardening Interests:

☐ Indoor (houseplant care/propagation)
☐ Outdoor (vegetable/flowers)
☐ Adaptive gardening tools
☐ Horticultural crafts
☐ Other:

Goals:

1.
2.
3.
4.

Discharge date: **City/State:**

Attending physician: **Med. Record #:**

Referred by: **Date of Referral:**

Horticulture Group Participation

Date: Activity:

Reviewed Videos:

Date:
_____*Intro to Square Foot Gardening*
_____*Square Foot Gardening, Vol. 1, 2, 3*
_____*How to Grow Cool Weather Veg.*
_____*Hydroponic Gardening*
_____*Other:*

HT Activities—In-House and Outings

Date: Activity:

Written Resources:

☐ *Raised Bed Gardening*
☐ *Adaptive Garden Equipment*
☐ Tool Catalogue Vendor List
☐ Garden Plot Referral
☐ Review *The Enabling Garden*
☐ Garden Design Illustrations
☐ Other:

Adaptive Equipment:

Wide U-cuff (S M L)
Forearm cuff (S M L)
Submitted request
Other

Comments/Recommendations:

D.C. Note:

(Contributed by S. Hall for Craig Hospital)

Appendix D:
Using Horticultural Therapy
for Mental Health

Horticultural Therapy Methods
© 2006 by The Haworth Press, Inc. All rights reserved.
doi:10.1300/5676_09

HT APPROACHES AND ACTIVITIES FOR MENTAL HEALTH

Diagnosis:	Possible Issues:	Modifications and Adaptations:
Major depression	Low motivation and energy, no desire to do anything for himself/herself	Suggest they pot up plant to give a family member
	Neglect of hygiene and grooming, lack of appetite, unhealthy level of sleep	Use caring for plants as metaphor for caring for self—what do each need to survive and thrive
	Low self-esteem	Use plant care to gain sense of accomplishment
	Safety issue: Suicidal ideation	Monitor all tools and supplies, use nontoxic plants, use plastic pots instead of terra cotta, use no sharp objects
Bipolar affective disorder	Rapid and pressured thoughts and speech, and high level of distractibility creates difficulty in listening to and following directions	Have them repeat directions, then monitor process so they don't skip steps or do steps out of order—may help to provide simple written list of steps
	Extreme irritability	Set clear, fair, consistent limits but stay calm. May need to separate irritable client from peers to prevent arguments
	Safety issue: Impaired judgment due to grandiosity and impulsivity	Supervise bipolar clients closely for safety purposes to prevent them from taking inappropriate risks or engaging in risky behavior
Schizophrenia and schizo-affective disorder	Poor social interaction due to paranoia and other symptoms provides little opportunity to nurture others	Plants provide nonthreatening way to connect with another living thing, and an opportunity to nurture. People with schizophrenia often show great patience with repetitive tasks such as weeding, watering, and plant grooming.
	Paranoia regarding unfamiliar people and objects	Use fragrant herbs to break down these barriers. Pick an herb leaf, tear it in half, and eat half to concretely show it is safe.

Diagnosis:	Possible Issues:	Modifications and Adaptations:
	Disorganized thought processing can lead to difficulty following and remembering directions, and to impaired safety judgment	Work side by side with client and perform one step at a time as they follow your lead. Use the same terminology and steps in the same order every time.
	Hypersensitivity to sensory stimuli which creates overwhelming distractions and restlessness	Reduce clutter, and limit choices. Use tasks that take small segments of time with chances for breaks to walk for a few moments to relieve restlessness.
	Extreme sensitivity to criticism, negativism, and disrespect	Use neutral tone of voice, and be aware of own body language—stay positive at all times.
Substance dependence and abuse	Irritability and restlessness due to constant craving for substance	Redirect attention to pleasurable task such as plant propagation or gardening to help clients learn more healthy ways to create pleasure, and to focus on something besides themselves
	Low frustration tolerance, impulsivity, and poor problem-solving lead to pattern of quitting activities when frustrated	Plant-care activities provide low-key and nonthreatening ways to focus on problem solving. Ask, "How can we work together to solve this?" when a problem arises. Offer support and information.
	Physical restlessness and anxiety	Allow frequent breaks or perform horticultural tasks requiring movement, such as gardening tasks
Anxiety disorders	Tremendous energy spent on avoiding situations in which they might experience anxiety or panic which they feel very intensely	Divert attention with horticultural activities that are interesting and self-motivating. Have clients frequently rate their level of emotions on scale of 0-10 to help them learn to modulate own feelings.

Diagnosis:	Possible Issues:	Modifications and Adaptations:
	Extreme physical discomfort from physiological symptoms of anxiety and panic	Distract attention from physical symptoms using relaxing effects of plants, nature, and gardening. Teach relaxation techniques using deep breathing, and meditative aspect of garden. Fragrant plants may be of special use. Allow client to release anxiety via physical activity.
Dementia	Safety issue: Varying levels of confusion and disorientation	Use nontoxic plants and watch carefully so clients don't ingest plants or soil. Keep watch on tools.
	Short-term memory impairment	Do activities one step at a time. Use old-fashioned plants to assist in reminiscence about plants/gardening in clients' youth. Use seasonal plants/activities to assist in orientation. Use multisensory approach.
	Safety issue: Impaired executive functioning and physical restlessness/agitation leads to increased fall risk	Make sure clients are seated safely for indoor and outdoor horticultural tasks. Allow pacing/walking in safely enclosed courtyard or use circular garden pathways that return to the building. Use path materials that are not slippery, especially when wet. Encourage clients to use walkers, canes, and other assistive devices correctly.
	Safety issue: Impaired executive functioning, impulsivity, and poor judgment can lead to combative and assaultive behavior—often with little or no warning	Watch where your body is in proximity to client. Don't become trapped in corner. Don't get close enough to be grabbed or hit. Keep hair pulled back so it can't be pulled. Eliminate necklaces and long earrings that can be pulled.
	Short attention span, and low frustration tolerance	Use plant activities that are simple and provide immediate gratification.

(Contributed by S. Sieradzki)

Appendix E:
Example of an Individual Program Plan

Horticultural Therapy Methods
© 2006 by The Haworth Press, Inc. All rights reserved.
doi:10.1300/5676_10

MOUNTAIN VALLEY DEVELOPMENTAL SERVICES
INDIVIDUAL PLAN

The following documents are examples of the documentation used for individual treatment planning for a person who participates in HT. The first document illustrates assessments and overall goals and plans. The following two show short-term HT goals and a charting system, respectively, for the same individual.

Mountain Valley Developmental Services **Date of Staffing:** **Name:**	
Medical Assessment: **DATE OF EXAMINATIONS:** Physical 03/25/04 Recheck 1 year Eye 05/26/04 Recheck 1 year Hearing 02/16/01 To be scheduled Dental 10/19/04 Recheck 1/25/05	Medical office—Continue to arrange appointments, and transportation to and from appointments; arrange and monitor that follow through is completed; instruct staff as needed for follow through. Follow up as directed from health care professionals.

CURRENT MEDICATIONS:
Staff may supervise/administer medications
 	
(1) Peroxide—3-4 drops in each ear everyday
(2) Atenolol 50 mg every day
(3) Multivitamin
(4) PRNs
    a Advil prn shoulder pain
    b. Pepto-Bismol per package prn diarrhea
    c. Tylenol prn for fever/pain
    d. Robitussin DM prn cough due to cold
    e. Sudafed prn allergies December-March
    f. Polysporin oint prn minor cuts/abrasions
	All staff—Provide medication supports as described in Medication Administration class: obtain, store, administer, track, monitor, and dispose of medication; seek help from medical office with any questions.

HEIGHT: 5'8" **CURRENT WEIGHT:** 142 **CURRENT B/P:** 147/89	Med. staff—Monitor weight and B/P; maintain current immunizations. Res. staff—Monitor _____'s blood pressure 1x per month at home.

IMMUNIZATIONS: (1) Flu Vaccine—10/26/04 (2) Diphtheria Tetanus—02/24/04 (3) Hepatitis B Antibody Positive	Med. staff—Ensure all immunizations stay current.
ALLERGIES: Citrus, animal hair, mold, yeast, chocolate, nuts	All staff—Observe allergies, treat with PRNs; report any concerns to MVDS medical ASAP when noticing any allergic type reactions.
RESTRICTIONS: Physical activities as able	
TREATMENTS: (1) Drink 3 glasses of water every day. (2) Liver function test every 6 months. (3) Physical therapy exercises 1x/week. (4) Trim toenails & fingernails 2x/month (5) Check blood pressure monthly at home. (6) Wear sunglasses with UV protection when outside in sun. (7) Wear athletic supporter when horseback riding. (8) Brush teeth 2x/day with staff supervision. (9) Listerine cool mint, rinse 2x/day. (10) Floss teeth 1x/day. (11) Apply sunscreen SPF 15 or > before going outside. (12) Regular diet—NO CAFFEINE	All staff—Be aware of treatments and their frequency. All staff—Encourage water consumption; ensure that ____ drinks at least 3 glasses of water per day.
THERAPEUTIC DIET: Regular with the exception of allergy foods, no caffeine	All staff—Follow diet.
DENTAL: Moderate gumline plaque, heavy plaque on upper left teeth. Needs help brushing that area	All staff—Be sure that _____ brushes after every meal. Monitor _____ to make sure that he is doing a thorough job. Verbal prompts to get _____ to floss may be required.

VISION: Myopia right and left eye. Wear sunglasses when outside w/ UV protection. No prescription required.	All staff—Encourage _____ to wear sunglasses as directed.
HEARING: Right ear with mild to severe mixed hearing loss. Left ear with moderate high frequency and mild low frequency hearing loss. Right and left ear infection treated and resolved by Dr. B. Continue hydrogen peroxide drops every day as ordered by Dr G. to prevent cerumen impactions. Follow up with Dr. G. PRN. Will follow up with Dr. B. if _____ experiences any ear pain.	All staff—Continue treatments for hearing as ordered.
SKIN PROBLEMS: (1) Apply sunscreen 15 SPF 30 minutes before going outside. (2) Lac-Hydrin 12% to calluses on feet **CURRENT DIAGNOSES:** (1) Moderate mental retardation (2) Cerebral palsy	All staff—Help _____ apply sunscreen as needed. Help remind him when it may be necessary to apply the lotion.
CURRENT MEDICAL CONCERNS: **PROBLEM:** Gingivitis with heavy plaque on teeth on cheek side **ASSESSMENT:** At _____'s last dental exam on 10/19/04, the hygienist reported that _____ had heavy plaque build up at the gumline especially on the upper left teeth with generalized bleeding. Plaque has been noted to be an issue at all visits in the last year. _____ is now on a recheck every 3 months.	**PLAN:** Staff to prompt _____ to bite teeth together and brush on cheek side to teeth and also on back teeth. Please supervise brushing at least 1x/day. Continue use of mouthwash and flossing as ordered. Continue 3-month follow-ups.

PROBLEM: Hearing loss **ASSESSMENT:** _____ has mild to moderate high frequency hearing loss in both ears, probably as a result of recurrent otitis media infections over the years, as reported in the yearly audiology report on 2/16/01.	**PLAN:** Continue with hydrogen peroxide drops daily as ordered. Seat _____ with preferential seating in a conversational situation with the left ear closest to the primary speaker. _____ should use ear protection when exposed to loud noises. Staff should notify the Medical Office if _____ experiences any symptoms of otitis media (ear infection), such as fever, c/o ear pain, drainage from the ear, or change in his normal pattern of behavior. As well, monitor for indications of further hearing loss.
PROBLEM: Prostatitis **ASSESSMENT:** Because _____ was voicing complaints about difficulty urinating, he was evaluated in Medical Clinic on 4/19/01 by Dr. B. who diagnosed him with prostatitis. Prostatitis is an inflammation of the prostate, which is the gland surrounding the neck of the bladder and the urethra, causing difficulty urinating. He was re-evaluated by Dr. B. on 5/17/01 at which time he reported the prostatitis as chronic. _____ was seen by Dr M. (urologist) for F/U of dysuria and found to have a residual volume in his bladder of 127cc. After his annual physical he was treated for presumed prostatitis.	**PLAN:** Follow-up with Dr. B. as necessary. Staff to monitor and report: • Urgency to urinate. • Frequent urination • Waking up at night to urinate. • Difficulty starting urination and emptying the bladder

PROBLEM: Transurethral resection of the bladder

ASSESSMENT: Dr. F. performed a transurethral resection of the bladder on _____ on 6/9/99 because a stricture in his bladder from scar tissue was causing massive urinary retention and edema to his lower extremities. Following the surgery, a suprapubic catheter was placed through his abdomen into his bladder, allowing the surgical area to heal. With the catheter in place, the bladder muscles were also strengthened allowing _____ to urinate normally when the catheter was removed. _____ had been seen for follow-up with Dr. F. and he reported that the problems with his bladder had been resolved

PLAN: Follow-up with Dr. F. as necessary. Staff to monitor _____ for signs of abdominal distention from a distended bladder and also to monitor for any swelling of his lower extremities or complaints of difficulty urinating. Staff to also monitor for signs and symptoms of urinary tract infection and report to the Medical Office:

- Frequent urination
- Painful urination
- Cloudy or foul smelling urine
- Burning on urination.

PROBLEM: Cerebral palsy and bicipital tendonitis

ASSESSMENT: Participates in equine riding every week. Posture, gait, and self-confidence continued to improve with equine riding. _____ also reported a decrease in muscle pain with equine riding. _____ was evaluated for orthotics by the orthotic specialist from Grand Junction and the physical therapist who felt that _____'s gait could be improved with the use of braces. The braces were constructed and _____ wore them intermittently on a trial basis; however, he was unable to tolerate them even with adjustments. His gait did not seem to improve and they only created much frustration for _____. It was then decided by the orthotic specialist and the physical therapist that the use of braces would not be pursued any further.

PLAN: Continue equine riding as ordered. Staff to continue to evaluate _____'s balance and report any falls or c/o muscle pain to the medical office.

PROBLEM: Hypertension

ASSESSMENT: _____ was placed on atenolol by for elevated blood pressure. His B/P is to be monitored monthly and _____ wanted to be notified if his B/P was above 150/90. B/P has been in a normal range since being placed on the atenolol.

PLAN: Staff to monitor and record B/P monthly. Dr B. to be notified if B/P above 150/90. Continue atenolol as ordered. Staff to limit the salt intake in diet. Follow-up with Dr. as necessary.

Residential and Vocational Assessment:

PERSONAL CARE:

_____ is very independent with his personal care. He does a good job in caring for his teeth, staff prompt him to brush, and help him to put new batteries in his toothbrush. Staff assists with mouthwash and flossing. _____ is capable in bathing himself, but staff do not leave him unattended for safety reasons. (_____ is unsteady in his gait.) Staff assist _____ in getting in and out of bathtub and assist him in drying off. _____ takes care of his own personal needs while at the agency.

HOME LIVING:

_____ lives in Rifle at Vista House (group home). _____ is capable of picking up after himself, dusting and vacuuming when asked. He also makes his bed independently & puts away his own laundry after it is washed and dried. _____ enjoys socializing with his friends at Vista House in the living room, his room, or outside on the deck. _____ likes to watch TV, videos, & DVD's. _____ went to Hawaii in November and had a great time. He shared pictures and other interesting souvenirs. _____ went to California to spend Christmas with his sister. He is very excited about this.

Res. staff—Continue to assist _____ with all of his hygiene needs as needed: bathing, shaving, toileting, tooth brushing, dressing and undressing, etc. Ensure that _____ is showering regularly and washing in all the correct areas. Continue to monitor his water temperature when showering.

Res. staff—Monitor _____ when brushing his teeth with verbal and physical prompts as necessary. _____ will continue to live at Vista House in Rifle, as this is the least restrictive environment for him at this time, and he enjoys living there. _____ requires 24 hours on-site supervision.

Continue to request that _____ does his own chores. Praise him on a job well done.

All staff—Encourage _____ to continue traveling as it allows him to grow in many ways.

Continue to administer all of _____'s meds and record daily, as he is not self-medicating.

MVDS Medical—Continue to be responsible for scheduling all medical appointments, transportation to and from the appointments, and any necessary follow-up.

ROUTINE DAILY HEALTH CARE (medical/dental) SPECIAL DIET:

Mountain Valley provides for _____'s health care by scheduling and transporting him to appointments. He does not receive any medications during the day. Vista staff administers all of _____'s medication. _____ is not very picky with his food, eats well, and often asks for seconds. He is on a low-sodium diet. MVDS nurses schedule all of _____'s medical appointments. _____ has a very high tolerance for pain but most times he will let staff or nurses know if he is hurting.

All staff—Be aware of _____'s low sodium diet and all allergies.

COMMUNITY AND HOME SAFETY SKILLS:

_____ responds well to fire drills he knows to get out at the nearest exit. _____ needs some staff supervision when in the community, e.g., when to cross the street (sometimes he gets in a hurry and just wants to go). _____ is very social with strangers and needs reminded that when he goes on his trips to be careful who he visits with. _____ does know his name, phone number and address but he is often difficult to understand at times. When he goes on trips it is noted on his ID card that he has on his neck keychain, in case of emergency please call #'s. _____ is always accompanied by staff when out in the community during the day. He has general safety skills, but needs reminders to look for cars.

All staff—Continue monthly safety drills in house and at day program.

All staff—_____ needs supervision while in the community due to limited safety skills.

All staff—Talk with _____ about whom he is and is not to speak with while out in the community. Be sure he has his ID card on him at all times.

All staff—Give _____ verbal praise for knowing his phone number and address, and continue to quiz him to make sure he retains the information. If either change ensure he learns the new information.

DECISION MAKING:

_____ is capable of making decisions on what to wear, daily routine, and activities that he wishes to attend. _____ is capable of choosing tasks he wants to do in the greenhouse and which activities he would like to participate in. Sometimes he may need encouragement to do an activity if he doesn't understand what is being asked of him or if it is new.

All staff—_____ is his own guardian and can make his own decisions. _____'s sister plays a very active role in his life and can assist _____ with any major decisions, as can his direct care staff, if asked by _____.

ACADEMIC SKILLS:

_____ is very good at writing his name and recognizing his address. He is able to write some words but cannot write complete sentences. He will ask staff if he does not know how to spell something, and is able to copy it. He loves to draw and do artwork.

_____ is capable of following simple verbal one-step directions. His attention span is short but of course slightly longer if he is doing something of interest to him. _____ knows that he can purchase items with money yet doesn't understand its value.

He is able to count and to recognize numbers, and can correctly weigh the herbs. _____ knows how to do a variety of tasks in the greenhouse and knows where equipment and supplies are kept. He may need to be asked several times if he wants to do a task before saying yes or no, taking time to process the question. Occasionally he will think he understands what he is to be doing, but will find out later that he didn't quite know, even if he said he did.

All staff—Continue to offer _____ the opportunity to expand his academic skills, helping him write letters and create artwork for friends and family. Work with _____ on things that will stimulate his learning skills and motivate him to think.

Res. staff—Assist_____ in money management and budgeting for desired items. Continue to work with _____ on counting and the value of money.

Voc. staff—Continue to give _____ activities that will challenge his academic skills and help him maintain his current skills. Allow _____ the time needed to process your requests of him. Question him to assure his understanding of the task he is being requested to do.

COMMUNICATION:

_____ is capable of communicating well verbally with his housemates and staff, however sometimes he is difficult to understand. _____ doesn't have a problem with repeating himself when someone does not understand. He will let staff know his wants and needs. He enjoys interacting with the other clients and communicates with them appropriately. He likes joking with the others as well as with staff.

All staff—Ask _____ to repeat as necessary to understand his needs/desires.

FINE/GROSS MOTOR:

_____ walks with an uneven gait, he drags his one foot. _____ needs staff reminders to slow down and be careful when he is in a rush. _____ is able to do most anything he wants to. _____ has adequate fine motor skills for the tasks in the greenhouse. His gross motor skills are more limited, but it does not interfere with his work. He is still able to do any task he chooses, and can get up and down from the ground. _____ rides horses at Sopris Therapy, which helps to maintain and improve his gross motor skills. He enjoys going and talks about it with the others who attend often.

All staff—Make sure reminders are given to _____ as needed to slow down/be careful.

VISION/HEARING:

_____ has some hearing loss in both ears. He does not require eyeglasses, but wears sunglasses when he is outside.

All staff—Remind_____ to wear sunglasses when outdoors.

All staff—Notify the medical office with any concerns regarding a change in _____'s vision or hearing.

SOCIALLY APPROPRIATE BEHAVIOR:

_____ usually has quite appropriate behaviors both at home and in the community. _____ gets along well with the other clients and staff. He shows concern toward them and enjoys joking around with everyone. He is friendly and usually in a good mood. ____ enjoys social activities here at the agency, as well as the dances and special events. There are times, however, when he can become upset at his peers and residential staff. He may yell, or rage about the house, or be excessively bossy with his peers. Occasionally, _____ may physically threaten others. _____ still has a tendency to become upset if someone redirects him when he doesn't want to be redirected, or if another client is doing something he doesn't think they should be doing. If he was asked to do something in a way he didn't like, he can become upset and will call the person "bossy." During these "problem" times, staff redirect _____ until he calms down. _____ tends to focus on tragic events and world tragedies. He may stay on these topics for hours if not redirected to a more positive topic.

All staff—Continue to support _____ with recognizing appropriate social interactions and behaviors.

All staff—Continue to redirect _____ when he begins to focus on tragic events from the past. Remind him these things happened a long time ago and ask him what is going on with him today.

COMMUNITY PARTICIPATION:

_____ likes to go out into the community to participate in various activities. These activities include: movies, dances, and bowling with the MVDS Recreation Department. Also, _____ likes to go out to eat on "Guys Night Out", and goes to church on Sunday. _____ enjoys fairly frequent vacations; he went to Hawaii with other MVDS clients on a Trips Incorporated vacation. _____ visited his sister in California for 2 weeks during the holidays. He loves to travel and loves to visit with his family. At the agency, _____ will occasionally go on an herb delivery or other greenhouse outing. His participation with Enrichment Services has become less frequent but he will sometimes still go with them on picnics, swimming, or to do art.

All staff—Continue to offer _____ a variety of activities throughout the day.

Res. staff—Continue assisting _____ in planning trips to see his family.

EMPLOYMENT:

_____ works in the greenhouse full time. Some tasks that _____ enjoys doing are cutting and bagging herbs, dried herbs, plant maintenance, and transplanting. He prefers working outside and will usually self-initiate tasks, such as sweeping, raking, shoveling snow, or weeding. _____ is generally cooperative and usually willing to try new tasks with staff assistance. He is self-motivated and will sometimes initiate cleaning tasks or work outside on his own.

_____ will continue to work in the greenhouse Monday through Thursday and participate in the Rifle program on Fridays.

All staff—_____loves working outside.

Please be sure that _____ is dressed appropriately for the weather conditions. _____ needs to wear work gloves with insulation in the winter months.

All staff—Verbally praise _____ for self-initiating tasks. _____ is a hard worker. Let him know what an asset he is to the greenhouse.

SPECIALIZED EQUIPMENT:
N/A

RESIDENTIAL CURRENT GOAL:

_____ will redirect from compulsive conversation about deaths and world tragedies (that happened many years ago) to current events/current happenings.

Continue Residential ISSP: _____ will redirect from compulsive conversations about deaths and world tragedies to current events/current happenings. Target: 8/05.

OBJECTIVE 1:

When _____ is speaking compulsively about past death/world tragedies he will redirect to more positive conversations about current events/happenings within two verbal prompts.

NOV. 90% OCT. 87% SEPT. 86% AUG 93% JULY 100% JUNE 100%

Criteria is 95% for 6 consecutive months—has not currently met criteria.

RECOMMENDATIONS:

- Continue current goal
- Continue living at Vista House
- Continue staff support in enabling _____ to participate in various community activities

Residential Coordinator,
L. J. M.

VOCATIONAL CURRENT GOAL:

_____'s goal is to increase his vocational skills by completing one task each day he is at the greenhouse. The past months percentages have been July—69%, August—94%, September—93%, October—100%, November—100%, and December—91%. He is doing well, but has not yet reached the criteria of 95% for 6 months.

Follow through on all residential recommendations.

Res. staff—Please stay in contact with _____'s sister L. regarding the wonderful packages she mails to _____. Let her know when they are delivered and that everything arrived in one piece.

D/C Vocational ISSP:

_____ will increase his vocational skills. 1/05.

Begin New Vocational ISSP:
_____ will work 3 hours in the greenhouse per day. Start: 2/05 Target: 8/05

Rights/Dispute Resolution/ Grievances/Complaints:

Directly after this IP, _____ and his case manager went over his rights packet. _____ is his own guardian and says that he understands his rights and the appeals process. He does require staff assistance with understanding the specifics of his rights. _____ chose to keep his rights packet. _____ was given choices at his IP and chose to actively participate. There are no concerns at this time for _____'s health, welfare, or safety.

MOUNTAIN VALLEY DEVELOPMENTAL SERVICES
INDIVIDUAL SERVICE SUPPORT PLAN (ISSP)

Person: A.
Program: Vocational
Beginning Date: 1/05
Actual Completion Date:
Completion Expected: 8/05

Short-Term Goal: A. will work up to 3 hours in the greenhouse per day
Objective 1: A. will produce 1 hour's worth of work.
Objective 2: A. will produce 2 hour's worth of work.
Objective 3: A. will produce 3 hour's worth of work.
Criterion: 80% for 6 consecutive months, objective one. Then A. will move to objective two and so on.

Prerequisite Skill: Desire to participate in more greenhouse activities, knowledge of tasks.

Current Performance: A. has difficulty staying on task. He will start working then he will easily become distracted and fall off task.

Purpose of the Program: To help A. start a work-related task and finish the task in a timely manner.

Materials Needed: Supplies and equipment as needed to do tasks, staff encouragement.

Person(s) Responsible: Greenhouse staff, A.

Timelines for Review:
- Program Manager: Weekly, Monthly
- Case Manager: Bi-yearly
- BPRC/HRC: n/a

Steps (task analysis/behavioral interactions):
1) Staff will give A. the tools necessary to accomplish the work presented to him.
2) Staff will help A. stay on task with verbal prompts.
3) Staff will help A. build his capacity to work for longer periods of time.

Reinforcement: Verbal praise, increased job skills, paycheck.

Correction Procedure: Verbal reminders.

Data Collection Procedure:
+ Task completed during day
− Task not completed
0 No opportunity

_____   _____
Signature/Date   Signature/Date

MONTHLY GOAL CHART

Consumer: Program:
Year:
Staff Responsible:
Short-Term Goal:
Criteria: Phase: Frequency:

Day	August	September	October	November	December
1	1 +	Hippo	XXXXX	4 +	
2	2 +	4 +	XXXXX	4 +	
3	2 +	XXXXX	Off	0.5 0	
4	.5 hippo	XXXXX	Off	.4 +	
5	4 +	XXXXX	Off	XXXXX	
6		4 +	Off	XXXXX	
7		3.88 +	Off	4 +	
8	4 +	0.5 +	XXXXX	3.0 +	
9	3 +	Off	XXXXX	2.0 +	
10	2+	XXXXX	Off	0.5 0	
11	1 hippo +	XXXXX	Off	XXXXX	
12	3 +	4 +	Off	XXXXX	
13		1.5 pool+	Off	XXXXX	
14		3 +	Off	4 +	
15	4 +	1 (0.5) hippo	XXXXX	4 +	
16	4 +	4 (2) +	XXXXX	4 +	
17	4 +	XXXXX	Off	Hippo 0	
18		XXXXX	Off		
19	Hippo +	4 (3) +	0		
20		4 (4) +	.5 hippo		
21		0.5 (2.0) +	4 +		
22	2.5 = +	0.5 hippo 0	XXXXX		
23	0 pool	0.5 test	XXXXX		
24	4 +	XXXXX	4 +		
25	4 +	XXXXX	4 +		
26	4 +	4 +	+		

Day	August	September	October	November	December
27		1 +	Hippo +		
28		4 (3.5) +	4 –		
29	4 +	0.5	XXXXX		
30	Pool	4 +	XXXXX		
31	4 +	XXXXXX	4 +		
Total days	100%				
Total pts					

Key: 0 = Not offered, + = Offered successfully, – = Offer refused
(Contributed by Brenda Scrimsher)

Index

Page numbers followed by the letter "e" indicate exhibits; those followed by the letter "f" indicate figures; and those followed by the letter "t" indicate tables.

Horticultural Therapy Methods
© 2006 by The Haworth Press, Inc. All rights reserved.
doi:10.1300/5676_11